SCIENCE ON THE HOME FRONT

JORDYNN JACK

Science on the Home Front

AMERICAN WOMEN SCIENTISTS IN WORLD WAR II

UNIVERSITY OF ILLINOIS PRESS
URBANA AND CHICAGO

Manufactured in the United States of America
1 2 3 4 5 C P 5 4 3 2 1
This book is printed on acid-free paper.

Library of Congress Cataloging-in-Publication Data
Jack, Jordynn
Science on the home front: American women scientists in World War II / Jordynn Jack.
p. cm.
Includes bibliographical references and index.
ISBN 978-0-252-03470-1 (cloth : acid-free paper) —
ISBN 978-0-252-07659-6 (pbk. : acid-free paper)
1. Women in science—United States—History—20th century. 2. Women scientists—United States—Intellectual life—20th century. 3. Feminism and science—United States—History—20th century. 4. Sex discrimination in science—United States—History—20th century.
I. Title.
Q130.J33 2009
508.2'0973—dc22 2008054548

To Marguerite Epp (191?–1978)
Professor of Microbiology,
the University of Saskatchewan

Contents

Acknowledgments

In 1924, my great-aunt, Marguerite Epp, escaped from revolutionary Russia, where her parents were murdered, and was taken in by an aunt and uncle who lived in a small Mennonite farming community in Saskatchewan. At a time when women from her ethnic and religious background rarely finished grade school, she somehow became Professor of Microbiology at the University of Saskatchewan. The stories I've heard about Great Aunt Marguerite and other strong women in my family have inspired me to tell the stories of other women like her.

Like the women scientists I feature in this book, my work, too, depends on a network of friends, advisors, and supporters. This book began when Cheryl Glenn encouraged me to explore Lydia J. Roberts's writings and speeches, and it grew into a dissertation and then a book manuscript with her encouragement. I thank Cheryl for her guidance and advice throughout this process. Jack Selzer's insights proved central in shifting the focus of the book toward genre when he asked whether the genres of scientific writing might be gendered, whether women scientists were really playing "a men's game." At the time, I did not have a good answer to this question; this book is, in a way, my answer. I also appreciate the useful advice and support I received from Stuart Selber, Carolyn Sachs, Susan Squier, Marika Seigel, Jessica Enoch, Jodie Nicotra, Stacey Sheriff, and the rest of the rhetoric community at Pennsylvania State University.

My colleagues at the University of North Carolina helped me bring this project to fruition. Jane Danielewicz, Sarah Dempsey, Rebecka Rutledge Fisher, Jennifer Ho, and Adrienne Lentz-Smith kindly provided feedback on chapter revisions. John McGowan, James Thompson, and my cohort at the Institute for the Arts and Humanities provided indispensable advice on the publication process. Patrick Charbonneau, Katie Rose Guest Pryal, and Mark Sheftall imparted indispensable editorial assistance. Among my colleagues at UNC, all of whom have been tremendously helpful, I'd like to thank in particular Todd Taylor, Daniel Anderson, Erika Lindemann, Bill Balthrop, Carole Blair, Christian Lundberg, Sarah Sharma, Mai Nguyen, Laura Halperin, and JoAnna Poblete-Cross.

Thanks are also due to Joan Catapano, Rebecca McNulty-Schreiber, Tad Ringo, and Nancy Albright from the University of Illinois Press. Londa Shiebinger and a second anonymous reviewer provided thoughtful suggestions and comments.

Financial support for this project came from a number of sources. I thank the Social Sciences and Humanities Research Council of Canada (SSHRC) for the opportunity to begin research on this project through a dissertation grant. At the University of North Carolina, I received funding through the Spray-Randleigh Fellowship, the University Research Council, and the Institute for the Arts and Humanities; I am deeply indebted to the generous donors who made this support possible. In addition, I am grateful for the semester teaching release made possible by the College of Arts and Sciences at the University of North Carolina.

I am also indebted to the hardworking archivists who helped locate materials for many of the women scientists featured in this book and arranged permissions to quote from these sources: Janice Parthree at the Pacific Northwest National Library, who arranged copies of many declassified documents from the U.S. Department of Energy Reading Room at Hanford; Susan Snyder, archivist at Bancroft Library of the University of California for materials from the Japanese Evacuation and Resettlement Study; Karen Klinkenberg, archivist at the University of Minnesota for materials from the Florence Laura Goodenough papers; and Janice Goldblum, archivist at the National Academy of Sciences, who arranged for me to view materials from the Food and Nutrition Board's Committee on Dietary Allowances.

Introduction

Today I was reading about Marie Curie:
she must have known she suffered   from radiation sickness
her body bombarded for years   by the element
she had purified
It seems she denied to the end
the source of the cataracts on her eyes
the cracked and suppurating skin   of her finger-ends
till she could no longer hold   a test-tube or a pencil

She died   a famous woman   denying
her wounds
denying
her wounds   came   from the same source as her power

—from "Power," Adrienne Rich

Adrienne Rich's poem about Marie Curie, "Power," offers a fitting opening for a book about women scientists who followed in Curie's footsteps in America during World War II. World War II is often considered a turning point in the history of scientific culture, ushering in a new era of scientific research that was closely linked to military and industrial developments. Although a number of well-known scientific discoveries and inventions were made during this period (radar and the atomic bomb come immediately to mind), many of the accomplishments women made during the war have been forgotten over time. Unlike Curie, whose fame rested in part upon scientific research that led to her death, numerous other women struggled to live up to the ideal of the "exceptional woman" that Curie represented. Despite their efforts, few of them have been remembered or honored in the way Curie has. Most of them have received little more than a mention or short biography in works on women in science to date. This book seeks to redress their erasure from history, and in doing so, to shed some light on the gendered culture of science.

The Manhattan Project stands as the greatest example of how scientific power could contribute to military success during World War II, but it is widely attributed to a small group of male scientists—Robert J. Oppenheimer, Enrico Fermi, and Leo Szilard to name a few. Although several hundred women worked on the Manhattan Project, their contributions have mostly been forgotten. Women also participated in the Japanese Evacuation and Resettlement Study, a social scientific enterprise meant to gather information about interned Japanese-Americans' experiences with displacement and mass migration; attempted to change women's underrepresentation in psychology; and developed the Recommended Dietary Allowances (RDAs). By delving into archives, journals, newspapers, and other sources, I use these women scientists' wartime writings—reports, articles, editorials, correspondence, and more—to illuminate both their struggles and their successes in the detail they deserve. In doing so, I identify how the culture and language of science during World War II in America shaped women's professional and rhetorical careers.

World War II is also an interesting period in the history of gender and science because the role of women within the new scientific order was ambiguous. On the one hand, World War II represents a moment in American history when women were widely encouraged to enter into scientific study, both to fill the void left by men leaving for war and to boost the scientific effort on the home front. On the other hand, historians generally agree that the war marked only a fleeting shift in gender ideology, hiring practices, and workplace realities for women in science and technology (Honey 2–3; Gluck 17; Milkman 1; Rossiter, *Women Scientists in America* II). Historians have shown that women hired for semiskilled positions in war plants—epitomized by the popular figure of "Rosie the Riveter"—often received only cursory technical training, were hired into subordinate and sex-segregated positions, and were often expected to leave their jobs once the war was over. But what about the scientifically trained women who also hoped to take advantage of the war to find new career opportunities? Like their industrial worker counterparts, women scientists found themselves mostly channeled into appropriately "feminine" areas of science or subordinate positions on wartime projects.

Although the popular media championed women's big break in science, the realities of scientific culture remained stubbornly conservative. Faced with a shortage of male scientific and technical workers, the United States government's Office of Scientific Research and Development and the War Manpower Commission looked to women as an untapped reservoir of scientific talent. Books and pamphlets advertised jobs in metallurgy, meteorology, chemistry, mathematics, geology, geography, pharmacology, and physics. Even the most male-dominated fields, such as engineering, shifted to accommodate wartime needs. The United States Civil Service Commission reframed engineering—traditionally one of the most male-dominated of all scientific fields—as suitable to women's skills and interests, noting that "Feminine aptitudes may be well adapted to engineering design, testing and inspection, research, preparation of plans and maps, and computation" (*Calling Women for Federal War Work, n.p.*). Women were urged to pursue training in the applied sciences and were promised exciting new career opportunities. Echoing the excitement generated by these reports, Evelyn Steele wrote in a 1943 book, *Wartime Opportunities for Women,* "Women are being offered scientific and engineering jobs where formerly men were preferred. Now is the time to consider your job in science and engineering. There are no limitations to your opportunities" (99–100).

Despite the fervent appeals for women to apply their scientific skills to the war effort, the culture of science was not significantly altered to accommodate them. Academic science remained stratified, with women mostly occupying lower status positions and concentrating in lower status fields. Prominent male scientists snapped up most of the high-status wartime positions early on (Rossiter, *Women Scientists in America* II 2). In fact, historian Margaret Rossiter claims that employers seemed more willing to hire untrained women in low-level positions as "aides" or "assistants" than they were to hire women who already had scientific training for higher positions (*Women Scientists in America* II 13). Overall, while the raw number of women entering into scientific careers increased during the war, the overall *proportion* of women in scientific fields did not increase appreciably, inching from 4.0 percent of the National Roster of Scientific and Specialized Personnel in 1941 to just 4.1 percent in 1945 (*Women Scientists in America* II 24). Without shifts in the culture of science, those who succeeded during this period did so largely by internalizing dominant scientific values and discourses, not by challenging them.

The case of women scientists during World War II is relevant in our current context because, in some ways, female scientists face a comparable situation today. Over sixty years after World War II, global scientific and technical interests also pose a challenge for the scientific workforce in the United States, and the need to boost the number of women in science has become a recurrent issue. For instance, a 2006 report of the National Academy of Sciences offers new justifications for increased training for women in the sciences: "To maintain its scientific and engineering leadership amid increasing economic and educational globalization, the United States must aggressively pursue the innovative capacity of *all* of its people—women and men" (Committee on Maximizing 1). As was the case in World War II, training women scientists seems to be imperative within a context of global competition.

Despite the urgent need for trained scientists, some individuals continue to question whether women, as a group, can fit into the established culture of science. Also in 2006, Harvard President Lawrence Summers offered a number of speculations in a speech he gave at the National Bureau of Economic Research (NBER) Conference on Diversifying the Science and Engineering Workforce, the most controversial of these being his suggestion that women might lack some of the innate ability required for scientific work. However, many of Summers's speculations actually were concerned with the structures of scientific institutions, which have been designed around a male standard. Summers mentions that the timeline for advancement in scientific careers does not accommodate a few years off to have children. He notes that "in fields where the average papers cited had been written nine months ago, women had a much harder time than in fields where the average thing cited had been written ten years ago" (*n.p.*). While Summers does not go so far as to suggest that these structures should be altered to accommodate women, he does suggest that the norms of scientific culture may work against many women, forcing adherence to a strict schedule of publication and advancement. What is needed, then, is renewed attention to the culture of science and how it might change to accommodate women's needs and interests.

This leads us to another problem that Rich alludes to in her poem—the price of success in science. Curie paid this price with her life, but other women scientists paid in other ways, at the expense of their values and beliefs or of success for other women scientists. In order to succeed in a scientific discipline—many of which were still hostile to women—a woman scientist would need to enact the dominant discourses of scientific practice. While these discourses sometimes brought them success, they also stifled alternative perspectives and ways of doing science. Taught to see themselves as "exceptions," these women did not necessarily help to open up opportunities for other women.

In recent years, feminist scholars have argued that simply increasing the number of women in scientific disciplines is not enough to make these disciplines truly open to women. As Londa Schiebinger explains,

"Because modern science is a product of hundreds of years of active shunning of women, the process of bringing women into science has required, and will continue to require, deep structural changes in the culture, methods, and content of science. Women should not be expected to succeed happily in an enterprise that at its origins was structured to exclude them" (*Has Feminism* 11). Instead, the culture of science needs to change in order to encourage women's success. Among the aspects of scientific culture that require modification, scholars have focused on education practices (Watts), employment discrimination and stratification (Rossiter), and the role of domestic arrangements (Schiebinger, *Has Feminism*), to name a few.

The nature of scientific language has also received some attention from feminist scholars. Suzanne Fleischman has argued that scientific language limits the kinds of critical and discursive moves women may want to perform. According to Fleischman, science can be understood as a "Religion of Knowledge," a religion that is theoretically open to anyone, male or female, "*But the acts—speech acts, rites, critical gestures—many women want to perform are not among the ritual acts of that religion.* Which is why women can't just appropriate the discursive attire of scientists/scholars and have the authority to say what *they* (we) want to say" (1003). Donna Haraway has also called attention to the importance of reworking scientific discourse to account for alternative perspectives—a process she describes as untangling the "sticky threads" of technoscientific discourse and forging new threads (*Modest Witness* 15). These scholars agree that scientific language must change in order to accommodate alternative perspectives.

To date, however, the focus on language within feminist science studies has remained mostly at the level of scientific style—terminology, metaphors, grammar (such as the passive voice), and gender-neutral language. Evelyn Fox Keller, Emily Martin, Londa Schiebinger, and many others have pointed to gendered and sexualized metaphors of scientific language (Martin; Keller, *Reflections*; Schiebinger, *Nature's Body*). Often, these studies conclude that scientists must be trained to identify and eliminate such metaphors from scientific language as evidence of bias. In other words, the assumption is that gendered metaphors produce bad science, science that is not value-free and therefore not objective.

Feminists have successfully lobbied against some of the most obvious forms of gender bias in language. Groups such as the Council of Biology Editors (CBE), the Institute of Electrical and Electronics Engineers (IEEE) and the American Psychological Association (APA) now issue guidelines for scientific writing that include suggestions for avoiding sexist language.

For instance, IEEE recommends that writers use generic titles (such as "chair" instead of "chairman") and the phrase "he or she" when writing in the third person (rather than "he" alone) (21). These linguistic changes both reflect and encourage trends toward greater inclusivity and toward the recognition of gender bias in scientific knowledge.

Yet, aside from this focus on metaphors and other sentence-level features, scholars have paid less attention to the broader rhetorical frameworks of scientific discourse, and how those frameworks might relate to feminist projects for science. I argue here that revising scientific language on the level of genres and discourses, not just the level of style, is necessary in order to enact such changes.

Examining scientific genres can help us to better understand the culture of science. As part of the mundane practices of science, genres work to reproduce power structures in ways that often escape our notice. The concept of genre is useful because it helps us to understand how typified rhetorical actions reproduce ideologies and power relations in scientific culture. In recent years, the concept of genre has moved out of strictly literary study, where it has generally been used to categorize types of literature—the novel or epic poem, for instance. In rhetorical scholarship, researchers have demonstrated that genres also include the different kinds of texts we encounter in everyday life, such as technical reports, software manuals, or newspaper editorials. Researchers have also shown that scientific disciplines have their own set of genres, and that these genres help to accomplish much of the work of science, whether oral (in laboratory interactions and conference presentations), or written (through grant proposals, articles, and literature reviews) (Bazerman, *Shaping Written Knowledge;* Ceccarelli; Gross; Myers).

Genres can be identified in part by the textual features they share. For example, scientific articles are generally organized around an IMRD format: Introduction, Methods, Results, Discussion. Yet, rhetorical research has shown that genres refer not only to the textual features of certain types of documents, but that they also index the disciplinary cultures within which writing occurs (Bazerman, Little, and Chavkin 456; Yates and Orlikowski, "Genre Systems: Structuring" 14–15). Carolyn Miller has argued that genres express a "typified rhetorical action," or a response to a recurring rhetorical situation ("Genre" 15).

The features of the scientific article, for instance, respond to the recurring rhetorical situation wherein a scientist or group of scientists must report new research findings to an audience of peers. This audience will adjudicate the research findings based on the soundness of the scholarship, the significance of the research, and so on. The contents and style of a scientific article reflect and reinforce this rhetorical situation. The tendency for scientists to write in the passive voice, for example, helps them make their research sound more objective than if it were written in the first person. In this way, the genre of the scientific article supports the power structures within scientific fields, where objectivity becomes a primary value and means of disciplining researchers who are considered outside of mainstream practice. Genres bear the imprints of scientific cultures. For this reason, focusing on the genres women scientists wrote provides one lens with which to examine the culture of scientific institutions during World War II.

The concept of genre also provides an effective tool for analyzing the situation of women scientists because it foregrounds the relationship between institutional or disciplinary settings and individuals' rhetorical acts (Berkenkotter 329). Scientific genres help to shape, in Mary Lay's terms, "what the [scientific] community believes is truth, what constitutes its authoritative knowledge about important subjects, and what evidence the community uses to sustain or alter its knowledge" as well as the status of competing knowledge systems within society more broadly (21). This relationship between texts and contexts is not neutral, however, but shaped along unequal lines. As Dorothy Winsor has shown, genres institutionalize and enable distributions of knowledge and power, often in ways that limit access to knowledge and power to a select group (151). Thus, scientific genres, as is the case with any genre, "locate or position individuals within the power relations of institutional activity" (Paré, "Genre and Identity" 59).

In scientific fields, some individuals—often lower-status technicians or research assistants—produce less prestigious genres, such as lab notes or reports, while others—often higher-status scientists—use those notes to produce higher-status genres, such as scientific articles. These individuals often occupy different spaces and times within the organization of the laboratory. For example, at Los Alamos Laboratory, where the design and construction of the atomic bomb was carried out, women and men were often segregated. Some women worked as "computers," compiling data needed for the Project; others helped to coordinate documents and information; others worked as laboratory assistants; and still others served in communications as telephone operators. Often, these women worked in rooms full of other women compiling data, and were supervised by a scientist (probably male). These women's tasks were broken into smaller components; physicist Richard Feynman describes how, when there was no IBM machine available to perform a certain set of calculations, they "set up this room with girls in it. Each one had a

Marchant. But *she* was the multiplier, and *she* was the adder, and this one cubed, and we had index cards, and all she did was cube this number and send it to the next one" (125). The archives of the Manhattan Project bear evidence of this: technical reports written by women can be hard to uncover, probably because most of the women produced raw data to be incorporated into reports written by higher-ups, mostly men.

Within scientific disciplines, not all genres are equally valued. The scientific article, in particular, is a privileged genre, one that confers credit, status, and prestige. As Bruno Latour and Steve Woolgar have demonstrated in *Laboratory Life,* the scientific article forms part of the "cycle of credit" in scientific disciplines. A successful scientific article may lead to recognition, which in turn may translate into grant money and new equipment, which can help the scientist produce more data for more scientific articles (200–201). Historically, this cycle of credit has excluded many women scientists, who lacked institutional positions and, possibly, were less comfortable with the agonistic rhetorical styles and masculine codes of competition required to participate (Nye 77–78). In the 1930s and 1940s, women scientists often did not have as many opportunities as their male counterparts to publish scientific research articles. Women in psychology were channeled into applied positions that required them to spend time administering tests, counseling patients, and so on. Because many of them worked in clinics and schools, not universities, they did not always have access to laboratories, materials, and equipment to conduct experimental work that led to scientific articles. Nutritionists and home economists were expected to publish brochures, popular articles, and other genres meant for public audiences. Accordingly, women scientists, as a group, were hampered from joining in the cycle of credit and reaping its rewards. The genres women did publish more often—such as textbooks, government bulletins, and so on—were granted less significance within the cycle of credit. These writings for public audiences were not considered central to the scientific research enterprise and were downgraded in status accordingly.

Genres are also significant from an epistemological perspective because they help to reinforce and reproduce "discourse regulations," which "determine what *can* and *cannot* be discussed, as well as what *might* and *must* be discussed" (Paré, "Discourse Regulations" 112). The concept of discourse is useful because it encapsulates not just linguistic features, but also broader cultural and institutional factors that shape scientific practice as well as the identities of scientists themselves.[1] Norman Fairclough writes that discourses play an active role in "constituting social subjects, social relations, and systems of knowledge and belief" (36). For instance, the discourse of "objectivity" shapes not only what counts as scientific knowledge, but the self-conception of scientists, who view themselves as engaged in the pursuit of truth divested of any particular values or bias. In this way, dominant discourses can be exceptionally productive—they can and do enable scientific facts to emerge as facts. Yet, dominant discourses also constrain both scientific knowledge and the available identities for scientists.

In what follows, I identify four dominant discourses that constrained women scientists' contributions and successes in the wartime period: objectivity, technical rationality, gender neutrality, and expertise. The discourses of objectivity and technical rationality led women scientists to obscure or downplay their own personal and ethical investments in their research. For example, women working on the Manhattan Project sometimes either stifled concerns about the dangers of radioactive materials (or were made to stifle those concerns) because the dominant discourse privileged production over safety. Likewise, anthropologists studying Japanese-American internment had to downplay their personal identifications with those they observed in the favor of "objective" field reports; in this way, the ethical implications of internment received less attention than they might have otherwise. Similarly, the discourse of gender neutrality required psychologists to stifle their concerns about the unequal role of women in the field. Meanwhile, discourses of expertise led nutritionists to reproduce a hierarchical system of knowledge that subordinated the contributions of nutrition workers, homemakers, and nonspecialists to those of a select group of experts. In the chapters that follow, I consider in further detail how these discourses shaped women scientists' experiences during World War II. These discourses are insidious because, while they ensure individual success for some women scientists, they ultimately cement unspoken assumptions about the neutrality of scientific discourses, genres, and ideologies. Ultimately, in order to succeed, women scientists adopted discourses that mask the ways in which genres support institutional power, silence alternative perspectives, and prevent women from identifying the discrimination they faced. These discourses have implications not only for women scientists, but for epistemology and ethics. That is, they also serve to support the epistemic authority of science while limiting consideration of its ethical and societal implications.

For this reason, I argue that removing material and cultural barriers to women's participation in science is not enough to ensure their success in scientific fields. Nor are surface changes to the language of science—such as removing gendered metaphors or using gender-neutral language—sufficient (although they are necessary). Rather, scientific genres and discourses must also be altered in order to allow for more equitable and accountable distribution of power and knowledge.

Women have historically written a wide range of genres, including many nontraditional and popular forms as well as more conventional scientific articles, reports, and literature reviews (Wells). Since the eighteenth century, women have excelled at writing genres that adapt science for public audiences, such as advice books, government bulletins, or textbooks (Rauch; Baym; Gates and Shteir; Sheffield; Gates). Popularizations of science were considered appropriate genres for women, while more strictly "scientific" genres (such as research articles) were not. For this reason, women scientists' roles and relationships within scientific institutions and disciplines have sometimes been different. Susan Wells has shown that women doctors developed strategies such as the heart history, a genre that not only facilitated a more open, collaborative relationship between doctor and patient, but also enabled female physicians to more effectively intervene in the patient's family life, morality, and reproductive behavior (32–34). Developing these new *genres* helped female doctors to reshape doctor-patient relationships and institutional contexts.

Changes in rhetorical genres can help to bring about broader changes in social organizations and the forms of knowledge they produce because genres and institutions are mutually constitutive (Bazerman, Little, and Chavkin 456). In the writing of women scientists during World War II, I find not only capitulation to the dominant discourses and genres, but also experimentation and innovation. These changes were not always successful in the immediate rhetorical context, but they nonetheless suggest alternative rhetorical practices that might help to bring about more equitable forms of scientific practice.

In this way, focusing on genres also opens up alternative scientific rhetorics that may be especially useful for future forms of scientific knowledge that enable more egalitarian arrangements of people, material resources, and knowledge production. Because they embed longstanding assumptions, values, and arrangements of time, energy, and power, genres create a certain kind of inertia. Yet, genres also provide the grounds for change because they evolve over time in unpredictable ways. Anthony Paré argues that opportunities for genres to change can emerge "when an event occurs that does not match the anticipated, socially construed exigence to which the genre responds; or, in a related situation, when the genre is stretched too wide, and its forms and actions are inappropriate or ill-suited to the occasion" ("Genre and Identity" 61). For this reason, we may also locate within extant genres the potential for alternative forms of writing and knowledge making.

Outline of Chapters

Chapter 1, "Women Psychologists Forecast Opportunity," explores public and private rhetorics that shaped women scientists' sense of appropriate stances toward women's place in psychology. In 1941, women organized the National Council of Women Psychologists to lobby for greater professional opportunities. Their actions sparked a series of articles assessing the nature of the field and predicting future opportunities for women. I analyze articles in three different genres. Alice Bryan and Edward Boring wrote a series of reports based on statistical analyses, Florence Laura Goodenough wrote an editorial based on expert opinions, and Georgene Seward wrote a review essay that drew upon historical and cultural evidence. The genres chosen, as well as the position of each writer with an institutional network, ultimately shaped these writers' sense of the situation for women in psychology and the possible solutions to their lower status. Seward was able to offer perspectives that accounted more effectively for women's underrepresentation in psychology, but only by eschewing a gender-neutral position.

The following chapters explore how genres disciplined appropriate scientific writing by female scientists in three more fields. In Chapter 2, "Women Anthropologists Study Japanese Internment," I examine how discourses of objectivity shaped writing by novice anthropologists and the structures that developed in the Japanese Evacuation and Resettlement Study (JERS), led by sociologist Dorothy Thomas. Thomas employed several field researchers, including graduate students Rosalie Hankey and Tamie Tsuchiyama, to provide "objective" accounts of camp activity. The genres Thomas used to facilitate this research effectively militated against the kinds of reflective writing Hankey and Tsuchiyama seemed to want to produce—writing that would enable them to reflect on their experiences in the internment camps and how those experiences shaped their observations.

Chapter 3, "Women Physicists on the Manhattan Project," considers how female physicists produced assessments of risk in keeping with dominant institutional values of speed, efficiency, and urgency. I examine how a discourse of "technical rationality" shaped declassified reports by two young physicists, Leona Marshall and Katharine Way, who were assigned to study safety aspects of the plutonium processing plant at

Hanford, Washington. In their memoranda and reports, Marshall and Way were required to strike a balance between the urgency of production and their concerns about radiation hazards. Under a regime of technical rationality in which boosting the speed of production was paramount, environmental and ethical considerations were peripheral, at best.

Chapter 4, "Women Nutritionists on the National Research Council," examines the extent to which female scientists could shift scientific genres and networks in order to enact alternative methodologies and approaches. This chapter examines how nutritionists and dietitians adopted a discourse of expertise as a response to the exigency of war. Amid the urgent need to improve American's nutrition to support the war, Lydia J. Roberts and Margaret Mead both conducted research programs under the auspices of the National Research Council. Roberts led the development of the first set of RDAs, while Mead led studies of food habits in cultural context. By positioning their work as part of the national defense program, they promoted the RDAs as an authoritative guideline for nutrition planning and outreach. In this way, Roberts and Mead bolstered the authority and expertise of nutrition research, but for the most part failed to acknowledge the expertise of lower-status women including the dietitians, extension agents, and American homemakers who would apply that knowledge. This chapter suggests that simply adding more women to a discipline may not necessarily redistribute power and expertise along more egalitarian lines.

Women who wanted to make their mark in science during the war confronted a network of rhetorical and cultural practices that inhibited the kinds of research women could perform, how they wrote about that research, and ultimately their ability to capitalize on wartime opportunities to advance professionally. Discourses of gender neutrality, objectivity, technical rationality, and expertise influenced women scientists' choices of research topics, constituted appropriate identities for women scientists (often identities that effaced their gender), dictated appropriate stances toward risk and ethical concerns, and limited the extent to which they could enact changes to encourage participatory science. Drawing on these examples from World War II, I conclude by outlining some possible changes in genres that might help to alter, or "regender," scientific institutions in ways that will make them more welcoming to women. However, gendered scientific genres limit not only women's professional success, but also scientific research more generally. Regendering scientific rhetoric can help produce better science, that is, science that more fully interrogates subjectivity, ethical imperatives, and relations of power.

1. *Women Psychologists Forecast Opportunity*

> Tradition had it that, when war came to a country, men should fight and women should wait. Within our time, however, war has become more than men fighting on the battle lines and women waiting—and weeping—at home. War today means action everywhere and from everybody.
>
> —Gladys Schwesinger, psychologist and cofounder of the National Council of Women Psychologists

During the first week of September, 1939, when war broke out in Europe, members of the American Association of Applied Psychologists (AAAP) met at Stanford University for the association's annual meeting (Schwesinger 298). One of the psychologists present, Gladys C. Schwesinger, recalls hearing exciting rumors that American psychologists would be recruited to help with war training programs (298). Yet these rumors were not formally addressed by the AAAP until the following year. Schwesinger recounts that at the next annual meeting, held at Pennsylvania State University, several women in the audience were shocked to hear that many male psychologists had already been recruited for wartime projects. Although women represented a third of the field, not one woman was named in connection with the war effort. The American Psychological Association (APA) and the AAAP had formed an emergency council in the fall of 1939 to prepare for the impending crisis of war. Composed entirely of men, this committee later became the Emergency Committee in Psychology (ECP), part of the National Research Council (Dallenbach 497). The ECP appointed subcommittees on issues such as industrial psychology, training courses in military psychology, and the role of psychologists in postwar rehabilitation (496–97).[1] The ECP and a second organization, the Office of Psychological Personnel (OPP), also helped to arrange wartime appointments for psychologists, many of them in the military. These organizations overlooked the female psychologists clamoring to contribute to the war effort, distributing the prime wartime jobs primarily to men.

Schwesinger describes the frustration she and other female psychologists encountered at the AAAP meetings in 1939, 1940, and 1941, when women's concerns were ignored. At the 1940 meeting, as Schwesinger recalls,

> Elder statesmen told of delegations formed, committees drawn up, responsibilities assumed during the summer by the various psychological associations. Names of individuals already in action poured out of the reporters' mouths rapidly while each member in the audience sat open-eared, alert for mention of a particular corner in the new endeavors into which he or she could fit his or her individual contribution. As the list of activities and persons rolled on, not a woman's name was mentioned, nor was any project reported in which women were to be given a part. No promise was held that the pattern would be altered to include them. (298)

According to Schwesinger, women objected to this exclusion, but their male counterparts scoffed at their protests: "Promptly we were told that our job was to keep the home fires burning, that tradition favored the services of men in time of war" (298). Finally, as a response to this situation, Schwesinger and several colleagues formed their own organization, the National Council of Women Psychologists (NCWP), to determine how women could contribute to the war effort.

The NCWP helped female psychologists to address women's role in the war effort, in psychology as a discipline, and in the postwar world. After forming the NCWP, its members published several articles in journals such as the *Psychological Bulletin* and the *Journal of Consulting Psychology.* The latter published ten articles written by Dorothy Baruch on issues related to child care in wartime. A special issue on the service of women psychologists to the war effort included articles by Alice Bryan on the role of community libraries in supporting civilian morale, Baruch on child care centers and children's mental health, Leila Martin on selective services, and Dorothy Stratton and Doris Springer on recruitment and training of women in the U.S. Coast Guard reserve (Bryan "Educating"; Baruch; Martin; Stratton and Springer). Given this attention to women's roles in psychology, the NCWP seemed poised to contribute to the war effort and to trigger changes in the discipline itself. Spokespersons like Schwesinger encouraged actions that would publicize women's low status in psychology. Yet, after the war, there were few noticeable changes in the discipline—neither in terms of women's overall status nor in terms of concrete initiatives to improve that status.

Some historians of psychology claim that female psychologists failed to capitalize on a key opportunity for feminist action. James H. Capshew and Alejandra C. Laszlo write that the wartime environment had "provided women with an opportunity to vocalize concerns about discrimination that had long gone unstated. Their ambivalence regarding this role, however, was expressed by limiting the scope of their work and in electing noncontroversial representatives," usually well-established women in the field who subscribed to a meritocratic ideal of scientific achievement and prestige (175). For this reason, they suggest, the war actually marked "a relative erosion" in the professional status of female psychologists (158). Although these assessments may be correct, they overlook the rhetorical strategies women did employ in their efforts to bring gender issues to the fore. In this chapter, I argue that the rhetorical practices used to address the "woman problem" in psychology prevented women from arguing their case effectively. More specifically, the genres of articles aimed at assessing and forecasting opportunities for women in psychology ironically served to limit and constrain the very opportunities they envisaged. Further, by using a discourse of gender neutrality, prominent psychologists naturalized the existing institutional structures and practices in their discipline and drew attention away from the gendered culture of psychology.

By *gender neutrality*, I refer to arguments that deny, overlook, or explain away women's underrepresentation as symptomatic of women's own failings or predelictions, rather than acknowledging systematic, institutional, and cultural inequalities. As Margaret A. Eisenhart and Elizabeth Finkel have argued, gender neutrality is problematic because it "supports women's participation in lower status science while simultaneously obscuring or contradicting workplace expectations that are especially troublesome for women" (34). Eisenhart and Finkel show that gender neutrality glosses over unequal professional conditions. For example, their study suggests that the scientific workplace is often described as gender neutral, even though women are often encouraged to reproduce traditional gender roles that subordinate women or position them in tasks requiring nurturing or assistance (181, 183). According to Gesa Kirsch, the discourse of gender neutrality has led women to "assume roles and qualities traditionally associated with male behavior: assuming authority, displaying knowledge, arguing forcefully, debating with conviction, scrutinizing and criticizing other people's work" (3). Yet a gender-neutral perspective denies that these activities are part of a male rhetorical tradition and rules out any discussion of gender as a variable in women's careers. Women scientists have often been led to support the supposed gender neutrality of science, because calling attention to sex discrimination would put them in an even more precarious position. I argue here that a gender-neutral discourse ultimately militates against greater participation for women in science. Rather than challenging the unequal nature of scientific institutions, a gender-neutral perspective upholds meritocratic ideals while eliding structural inequalities in scientific institutions.

Gender and the Culture of Psychology

During the years leading up to World War II, psychology was dominated by men, especially in academia.[2] Although women comprised about a third of the field, they were systematically channeled into lower-prestige, service-oriented positions in clinics, schools, and hospitals. According to a 1941 study by Donald Marquis, women were overrepresented in schools (where they constituted 52.8 percent of psychologists) and guidance centers or clinics (where they made up 53.4 percent of psychologists) (472).[3] Yet they were underrepresented at in the academy, where women made up only 19.7 percent of faculty members (472). Moreover, those women who did work in academic settings tended to do so at women's colleges or clinics linked to psychology or education departments at large universities (Capshew and Laszlo 160). Women often occupied different professional spaces than did men, and had less access to material resources. This stratification was not simply a matter of personal choice, but also a matter of disciplinary practices and advice. In 1943, the APA president, John F. Anderson, recommended that, instead of changing hiring practices, psychology departments should accept only average women students, who would be happier working with children, rather than those of exceptional ability who would expect better opportunities (Walsh 20). Women were channeled into the applied areas of psychology that garnered less prestige and provided fewer resources to encourage productivity.

These occupational differences meant that women produced different genres as well. Women who worked in guidance clinics and schools often spent their time administering and scoring psychological tests, so their research often reflected this experience. They also developed new psychological tests, such as Molly Harrower's Group Rorschach Test or Florence

Goodenough's Minnesota Preschool Scale. Published articles on testing or administration seldom yielded the same kind of professional recognition as articles about broader theoretical issues or experimental results.

Despite inequalities such as these, the assumption that science functions as a meritocracy has pressured women scientists to deny gender as a factor in their professional output (or lack thereof). As Evelyn Fox Keller notes:

> Because they are "inside," they [women scientists] have everything to lose by a demarcation along the lines of sex that has historically only worked to exclude them. And precisely because they are rarely quite fully inside, more commonly somewhere near the edge, the threat of such exclusion looms particularly ominously. At the same time, as scientists, they have a vested interest in defending a traditional view of science—perhaps, because of the relative insecurity of their status, even more fiercely than their relatively more secure male colleagues. ("Gender/Science" 240)

In other words, if women scientists were to suggest that gender influences opportunity or success in science, they would be calling into question a fundamental value of science. To do so would further jeopardize their already tenuous position within scientific disciplines. Therefore, many women (like most men in psychology) internalized a gender-neutral position, leading them to value progress and meritocracy and to disregard the inequalities women faced in the field.

This gender-neutral rhetoric appears in three articles published in 1944 by Alice Bryan and Edward Boring, Florence Goodenough, and Georgene Seward. Their articles addressed the wartime situation for women in psychology, the reasons for their relative lack of advancement, and/or opportunities for women in the postwar period. Yet in all but one case, the writers in question espoused a gender-neutral rhetoric, one that limited the range of opportunities women could be expected to fulfill, and failed to address the need for change in the culture of science. Each of these articles fits into a different genre: Bryan and Boring wrote a statistical report, Goodenough wrote an editorial, and Seward wrote a literature review. As I will show below, it is not simply that the arguments differ in these articles, but also that the genre of the article itself circumscribed the kinds of arguments that emerged.

Alice Bryan and Edward Boring: Statistical Report

In 1944, Bryan, a member of the NCWP, collaborated with Boring, a prominent and somewhat conservative psychologist, to research the status of women in psychology. During the early years of the war, Boring and Bryan had worked together on the ECP's Subcommittee on Survey and Planning, and Bryan's repeated critique of women's lack of representation in the APA stimulated Boring's interest in the issue (Capshew and Laszlo 171). Boring asked Bryan to study the problem with him, and the pair published a series of three articles on the status of American women in psychology between 1944 and 1946: "Women in American Psychology: Prolegomenon" (published in the *Psychological Bulletin* in 1944), "Women in American Psychology: Statistics from the OPP Questionnaire" and "Women in American Psychology: Factors Affecting Their Professional Career" (both published in the *American Psychologist* in 1946). These articles use a quantitative and statistical approach to identify the progress women psychologists had made, their standing in the field, and potential reasons behind their underrepresentation in the upper echelons of psychology. The statistical report the authors used ultimately encouraged a gender-neutral approach, constraining the transformative potential of their work.

In their first article, Bryan and Boring briefly outline the problem they are addressing: "Something ought, we think, to be done about women in American psychology. We ought to know the facts about what they are doing, and we ought then to consider whether they have any special role in the American scene, what their special functions are and why, what aspiration would best represent the full utilization of America's women in psychology. In this first paper we undertake merely to discover the numerical facts" (447). This passage sets up the "woman problem" as an empirical issue; it assumes that knowing the facts about women in psychology could lead to proper solutions. Bryan and Boring outline a course of action based first on gathering the facts before moving on to deliberative questions about policy or procedure. Yet, these deliberative questions are already framed here by an assumption that identifying the *present* facts provides the best way to chart a course for the *future.*

Bryan and Boring's article best fits the genre of an empirical report, a type of "forensic" rhetoric, or rhetoric that is directed toward past events (Aristotle 32). As a genre, the empirical report generally limits its focus to quantitative information (Rude 76). However, as I will show below, the empirical report does not provide an effective basis for imagining future courses of action; it is not a deliberative genre and does not effectively address problems of feasibility, comparison, or cause and effect (Rude 77–78). As Ruth Oldenziel has written about engineering, "An exclusive attention to [statistical] figures tends to blame women for their inadequate socialization and to ignore the professional politics behind the creation of such statistics" (12–13). Similarly, attending to the statistics about women in psychology does little to account for the fuzzier factors of socialization, discrimination, and scientific culture that might have accounted for those statistics. For this reason, Bryan and Boring's report upheld the institutional structures of psychology, assuming that the structures themselves were gender neutral and that raising women's status in the field was simply a numbers game.

The empirical genre of the report focuses Bryan and Boring's attention on the existing facts for women in psychology. The opening paragraphs are peppered with factual claims, such as the following:

> Everyone knows what has happened. American psychology in the 1880's and 1890's got its foot in the doors of American colleges and universities [. . .] The higher education of women had lagged behind the higher education of men; and there were fewer women's colleges than men's colleges, fewer women students than men. Men disliked being taught by women more than women, accustomed to being led by men, disliked being taught by men. There just was not, at the beginning of American psychology—nor is there now—as much chance for a woman to get a job as a teacher of psychology as there was for a man. (447)

While these statements may certainly have been true, they give an impression of inevitability. Phrases like "Everyone knows" make it less likely for a reader to object to the claims, while the phrase "There just was not" makes the facts seem inescapable. Later on, Bryan and Boring state even more baldly that women's roles in psychology have been secondary to those of men: "The fact is that the activities of women in American psychology have never, taken in mass, been as noteworthy as the activities of men" ("Women in American Psychology: Prolegomenon" 448). Once again, the tone here is one of resignation to the existing state of affairs.

Next, Bryan and Boring turn to the statistics, based on their analysis of membership in the APA and the AAAP (449–51). They determined the number of women and men in each association and in each rank. Both associations had a two-tiered system, with higher ranking "members" or "fellows" and lower-ranking "associates"; in each case, women were more strongly represented at the lower levels of membership (450–51). In the APA, women made up 30 percent of the fellows, while in the AAAP, women made up just 20 percent of the full members (450). Bryan and Boring also count administrative appointments as "The most obvious way of assessing the activity of women in these two associations" (451), noting that once again women were underrepresented. Yet Bryan and Boring do not question the two-tier structure of these organizations—

a structure that would inevitably subordinate some individuals, most likely those who (like many women in psychology) worked in non-academic settings. Nor do Bryan and Boring address how cronyism or nepotism might have led to a concentration of men in administrative positions. Instead, they give the impression that this two-tier structure is natural and gender neutral.

This is not to say that Bryan and Boring take a purely conventional approach to the issue. It seems likely that Bryan, as a member of the NCWP, was concerned that women were not equally represented in the APA and AAAP. The authors note that while there had been only one female chair of the APA, and no female presidents of the AAAP, "The people who do the hard work—the secretaries of the section—are apt to be women" (452–53). Furthermore, they point out that many of the committees in both the APA and the AAAP involve issues and activities "that are traditionally considered male" (454), and that this might explain women's lower representation among them. For the most part, however, this section frames the issue of "women in psychology" as primarily a matter of representation within the existing system. The statistics, tables, and graphs included in the article indicate that women are underrepresented and that their representation should increase, but do not suggest that institutional structures should change to value the kinds of work women were already doing.

Bryan and Boring devote less attention to the reasons behind women's underrepresentation in psychology. For instance, they suggest that one of the reasons may have to do with the more general issue of women's underrepresentation in science, but they back away from the question of why that is the case: "As to whether the exclusion of women from wide participation in the established sciences is due to nature or nurture, we venture no opinion, but certain it is that modern civilization tends not to place them in 'science,' not in any large numbers" (453). They move on to a similar statistical analysis of the proportion of women in the sciences more generally, based on their analysis of the publication *American Men of Science* and the composition of the National Academy of Sciences, and they conclude that "Women ordinarily are not thought of as good scientists—not, at any rate, by most men scientists" (453). Statements like these continue the factual mode Bryan and Boring established early on, a mode that tends to accept rather than challenge women's inferior status in the discipline.

Rather than turning toward concrete suggestions to ameliorate the problem, Bryan and Boring conclude their article in a utopian mode: "If some superhuman decree, issued by an experimentalist-dictator, could arbitrarily reverse the number of men and women in important scientific positions without altering human nature, would that new imbalance, once established, tend to perpetuate itself? Or would some natural law of sex-difference presently assert itself and bring the proportions back to what they are now? We hope to have something to say about this matter in a later paper" (454). Here again, Bryan and Boring presuppose certain unchangeable facts—"human nature," "natural law of sex-difference"—and also suggest that only some kind of "superhuman decree" might change this situation. Their gender neutrality does not extend to assumptions about men's and women's abilities. Instead, it refers to the structure of the discipline of psychology, one that they assume to be neutral rather than shaped by gender relations. Overall, their article provides mainly description of the status quo, rather than analysis of the unequal institutional structures that led to women's underrepresentation.

The next two articles, while promising more attention to deliberative questions, mainly deliver more empirical data. In "Women in American Psychology: Factors Affecting Their Professional Careers" (1946), Bryan and Boring present the results of a detailed survey conducted of equal numbers of men and women working in psychology. The pair asked respondents how long they spent reading professional literature, conducting research, and writing each week. Their results point to some of the different ways in which men and women were situated within the institutional structures of psychology, especially in terms of its temporal structures. For instance, men spent more time on these activities than women did, whether they were married or unmarried. For these activities, married men spent an average of 6.9 hours a week, as opposed to 5.4 hours for married women and 6.2 hours for unmarried women.[4] Bryan and Boring find the results for married women somewhat predictable, given that "reading would suffer when pressure for time is great" ("Women in American Psychology: Factors" 16), but found the difference for unmarried women remarkable. Bryan and Boring posit that this difference may be due to gendered divisions of labor within the field, suggesting that "It seems possible to us that we are here again in the presence of a basic difference, that 'women's work' in psychology involves less time spent on technical reading than does 'men's work.'"(16). Certainly, women's work in clinical settings required different types of tasks, including counseling patients and administering tests. Yet, Bryan and Boring do not seem to question this statistic and what it might say about the field of psychology, nor do they speculate about any self-reporting bias in the numbers more generally. By calling this a "basic difference," Bryan and Boring suggest that it is static, natural, even unchangeable, a simple reflection of men's and women's preferences in psychological careers. They do not suggest that women might have been channeled into the less prestigious, applied forms of psychology rather than the more prestigious, academic jobs. In this way, they support rather than challenge the status quo.

They reveal similar findings in "Women in American Psychology: Statistics from the OPP Questionnaire." Bryan and Boring found that men spent more of their time teaching (36.9 percent of men's time versus 22.7 percent for women) and conducting research (15.5 percent of men's time and 10.2 percent of women's) than women did. Women spent more time than men did on counseling and interviewing (20.4 percent of women's time versus 13.8 percent of men's), and constructing and administering tests (24.3 percent of women's time versus 8.8 percent of men's) ("Women in American Psychology: Statistics" 76). In the previous article, they had already noted striking results for time spent on research and writing. Although married and unmarried women both spent an average of 3.3 hours a week on these activities, men spent 6.3 hours, on average, writing and conducting professional research ("Women in American Psychology: Factors" 17).

The discrepancies between men and women that emerge in these reports are perhaps not surprising. What is notable in these reports is that Bryan and Boring interpret these statistics as a matter of innate ability or personal choice, not as a matter of institutional structure or sex discrimination within psychology. For example, to explain why men tended to publish more than women, they suggest that women may be more likely to "assimilate without as great an urge to do something about what they learn" ("Women in American Psychology: Factors" 17). To account for women's lesser research output, Bryan and Boring also explain "that women on the whole are less prone to generalization than are men, that their work is often clinical or particularistic, that research is usually directed at generalization and that men, therefore, become more readily involved in it" (17). While one of their studies found that only 72 percent of women's total work time was spent on paid work (versus 98 percent for men), the authors argue that "Much of their unpaid work is doubtless socially useful (as housewife and mother) and may even depend on psychological training"("Women in American Psychology: Statistics" 79). The authors do not make the connection between women's greater time spent on housework and family activities and their lesser time spent on research or publishing, although it is not hard to guess that women charged with a disproportionate share of domestic tasks would have less time for professional activities. According to Bryan and Boring, women's lesser research output has to do with aptitude and dedication, not to gendered allocations of jobs or to other institutional factors. They do not claim that research interests or abilities are distributed equally among the sexes, but they do assume that the institutional structures of psychology are themselves neutral. In this way, the blame for women's lack of status is placed squarely on their own shoulders.

Bryan and Boring work rhetorically to reinforce the status quo, perhaps as a way to assuage the more conservative members of their audience. For instance, they suggest that most women psychologists had "accepted the cultural pattern"—they actually preferred the kinds of employment they have and were not particularly interested in more equal disciplinary arrangements ("Women in American Psychology: Statistics" 76). Further, they suggest that unequal pay between male and female psychologists—women made between 20 and 40 percent less than men—"derives, not from the fact that *women* are paid less than men in psychological work, but from the fact that *women's work* in psychology is less well paid than men's work" ("Women in American Psychology: Factors" 11). Here, Bryan and Boring assume that "men's work" and "women's work" are natural distinctions or choices made by individuals, so the division of labor does not reflect gender discrimination.

On matters of a procedural or deliberative nature, Bryan and Boring are remarkably conservative. They portray changes in the culture of psychology and the culture at large as untimely. Citing results from survey questions on job satisfaction, they claim that "it is clear that the women who find satisfaction outside the profession, in marriage and the home, are performing a social service by their contentment. The culture is changing slowly in these respects, and too rapid a change would create disequilibrium in unexpected places" ("Women in American Psychology: Factors" 13). In other words, while social evolution in sex roles was considered acceptable, some unmentioned disruptions would occur if these roles were to change too quickly. The authors do not seem willing to recommend shifts in sex roles within marriage and the family, or within the discipline. Instead, they present the hypothesis that "men psychologists are, on the whole, more strongly motivated to undertake research and writing," which leads, in turn, to renown and financial compensation within the field of psychology (20). Even though men's greater productivity appears to result from a greater amount of time dedicated to research, a factor that is at least partially institutionally and culturally determined, it is explained as a result of greater individual motivation. In this way, Bryan and Boring imply that women psychologists could gain greater professional status simply by working harder and accommodating the culture of psychology.

Similar conclusions appear in "Women in American Psychology: Statistics," where the authors write the following:

> The dissolution of sex prejudice will come eventually when a person is compensated for the actual value of his services, irrespective of his sex, race or religion. Sometimes sex, however, makes a true difference in value. For instance, interest in a job is worth purchasing, and there are sex differences in interests. It is doubtful, however, if anyone would be bold enough to lay down equivalences in social value between university research, college teaching, clinical guidance and industrial consultation. The levels of compensation in such diverse fields get themselves established by tradition in the culture. They may get changed gradually by the efforts of those who believe they are wrong and who can convince responsible authority that they are wrong. (79)

Note that the authors do not end with a call to readers to undertake this persuasive work or to hurry along this process of cultural change. Instead, the authors make this change seem inevitably slow, thereby decreasing the audience's sense of responsibility for change. Rather than viewing the war as an opportunity for women to push forward, they portray it as a moment without opportunity (or akairos, in rhetorical terms).

In this way, Bryan and Boring's reports yield to a gendered scientific culture that systematically disadvantaged women and devalued their work. Instead of directing attention to how that structure might change, Bryan and Boring could only suggest that women work harder to accommodate that structure. Bryan and Boring's statistical analysis supports a narrative of gender neutrality, one that assumes that the discipline of psychology itself does not discriminate, but simply that differences in men's and women's professional success lie in personal preference, effort, or inborn talents. Their quantitative approach contributed to this argument, because the questions they asked did not reveal the more subtle institutional factors that channeled women into less prestigious specialties, nor could it envision alternatives to the gendered division of labor that prevented women's advancement.

Florence Laura Goodenough: Editorial

In 1944, Florence Goodenough published an editorial calling on women psychologists to work harder in order to gain professional prestige. "We sometimes speak of opportunities as 'given' or 'offered,'" she writes. "In the scientific world, however, opportunities rarely come fortuitously. They must be created" (707). In her article, "Expanding Opportunities for Women Psychologists in the Post-War Period of Civil and Military

Reorganization," published in the *Psychological Bulletin,* Goodenough assesses the range of opportunities that might be available to women following the war. Unlike the statistical report Bryan and Boring favored, the editorial genre seems to encourage different forms of evidence, including expert opinion and personal anecdotes. Presumably, this genre offers opportunities for Goodenough to transcend the discourse of gender neutrality by offering reflections on the unequal nature of the institutions in which women worked.

But this is not what happens in Goodenough's editorial. Goodenough appeals to two separate audiences: the conservative male and female psychologists of her generation, upon whose goodwill Goodenough herself depended, and the more radical, younger female psychologists who hoped the NCWP would address the discrimination women faced in the field. To ensure the continued support of the first audience, Goodenough could not afford to make any overt arguments advocating institutional changes. Thus, Goodenough urges her female counterparts to work harder and publish more research in order to increase their standing in the field. As a concession to the second of these audiences, the younger female psychologists urging change, Goodenough suggests simply that they carefully identify niches in which they could flourish. Goodenough naturalizes the institutional structures of psychology as providing plenty of opportunities for both men and women, even as she recognizes the gendered division of labor in her field. In this way, her editorial most closely resembles the epideictic (or celebratory) genre in rhetoric, one that is responsible for assigning praise and blame (Aristotle 32). While the forensic genre focuses primarily on past actions, the epideictic genre is mostly concerned with the present state of affairs (even if, as Aristotle suggests, a rhetor using the epideictic genre may "find it useful also to recall the past and to make guesses at the future" [32]). In the case of Goodenough's editorial, women psychologists themselves receive most of the blame for failing to live up to the high standards set for academic success, while the discipline of psychology itself receives praise for upholding these high standards, which are assumed to be gender neutral.

A short description of Goodenough's personal history might indicate why she espoused this approach. Goodenough started her career as a schoolteacher, but earned her Ph.D. in 1924 at the age of 40 and accepted a position at the University of Minnesota (Thompson 126). Although she began her academic career late, by World War II, Goodenough had established a formidable reputation. She published extensively on child development, emotions in children, intelligence testing, and research methods. Because of her stellar reputation and moderate views, she was elected president of the NCWP in 1942 (Capshew and Laszlo 165). As one of a small number of "exceptional women" who had a strong reputation in psychology, Goodenough would have been unlikely to jeopardize her own position by advocating strenuously for other female psychologists.

Indeed, Goodenough felt ambivalent about the role of the NCWP and was not very interested in women's issues. Capshew and Laszlo note that she "was unwilling or incapable of recognizing a sex bias" (165), and that she frequently insisted, "I am a psychologist, not a *woman* psychologist" (165). In Adrienne Rich's terms, Goodenough seems to have been an "antifeminist woman," one whose accomplishments led her to consider herself free from the constraints of gender, "even though this freedom requires a masculine approach to social dynamics, to competition with others, to the very existence of other human beings and their needs (which are seen as threatening)" (82). In order to succeed in a competitive academic world, Goodenough had to internalize dominant scientific values. Thus, it is probably not surprising that Goodenough espouses a gender-neutral position in her editorial.

In the first two pages, Goodenough argues that women psychologists must make a greater effort to engage in scientific research. She opens by placing the problem squarely on the shoulders of women psychologists, blaming them for taking "a less active part than men in advancing psychology as a science and as a profession" (706). She goes on to claim that "the contributions of women to basic psychological theory have been few" and that "[t]he majority of women have shown little aptitude for devising research techniques of wide applicability, for formulating new problems or searching out new approaches to old ones" (706). This passage would have appealed to the more conservative members of Goodenough's audience.

Like Bryan and Boring, Goodenough also mentions women's lesser achievements in terms of publishing. She notes that women tend to publish less than men, which makes it difficult for them to achieve prominence in the field: "the fact that women are less productive than men means that they stand less chance of being elected to office because they are less well known [. . .] There are and have been in the past a good many notable exceptions to the general rule. But on average the difference exists, and its consequences are both real and inescapable" (707). Goodenough blames women for failing to live up to the high standards set for academic psychologists.

At this point, Goodenough outlines her own theory of opportunity, arguing that women psychologists must be savvy in identifying research projects:

> I believe it is not unfair to say that if, in the past, women psychologists have been accorded somewhat less recognition than men, at least a part of the differences may be ascribed to less outstanding accomplishment. If opportunities for women in psychology are to expand the impelling force must come from within. Women must train themselves to see beyond the little routine problem with which they have too often been preoccupied in the past and concern themselves with larger issues. And they must cease to use sex discrimination as an excuse for their failure to do this. There is not, and never has been, any sex barrier to thought. (707–8)

In other words, Goodenough rejects the notion that women had not been offered appropriate opportunities for research. If women have been less successful than men, she states, it is because they have failed to create opportune moments (which rhetoricians would term moments of *kairos*) for themselves, or to identify areas of research in which they could excel. Such an argument would surely appeal to the elite group of prominent psychologists Goodenough addressed, even as it chastised those female psychologists who made up the NCWP for failing to create such opportunities.

In this way, Goodenough denies that the gendered culture of psychology prevented many women from doing theoretical or empirical research. For example, she ignores the fact that women who worked in schools or guidance clinics might be more likely to work on "routine problems" because they dealt with those problems—such as administering and scoring psychological tests—on a regular basis. For those women, practical developments to streamline tests might seem like important work, even if it was not valued by the broader academic community, which valued more generalized, theoretical work.

Clearly, Goodenough's argument also ignores systematic discrimination against women in psychology, including the devaluation of the kinds of work women were encouraged to perform (such as clinical work) and the preference for male candidates for faculty positions. The types of positions women were likely to occupy also provided less time for research or professional reading, as Bryan and Boring's studies demonstrated. In order to gain status, by male standards, women psychologists needed academic positions that would provide them with the time and resources needed to publish, as well as a discipline that valued research in areas such as applied psychology and women's and children's psychology. Nonetheless, Goodenough's advice would have been persuasive to conservative male and female psychologists who supported the "bootstraps" argument that women simply needed to work harder to get ahead.

In the remainder of her article, Goodenough identifies further opportunities for psychological research that women could perform following the war. She points to potential openings in the field—something every scientist must pay attention to, as Carolyn Miller has suggested ("Kairos" 313). By beginning research on those topics in 1944, female psychologists could position themselves favorably to take advantage of postwar opportunities. Nonetheless, for the most part, the types of openings Goodenough describes are clearly gendered. For example, Goodenough suggests that positions in educational guidance and counseling at universities and colleges would probably increase alongside an influx of returning students after the war (709). Similarly, Goodenough cites statistics showing that the population of the United States, as a whole, was aging, and suggests that research and service in gerontology might provide opportunities for women, especially "if there is justification for the popular belief that women are more patient and sympathetic in dealing with handicapped people than men are likely to be" (710). Among the other opportunities she highlights are counseling women who had been industrial workers during the war but were required to give up their positions afterward; advising wives and mothers to help them adjust to life after their husbands returned from war; and researching war marriages, the rise in divorce and illegitimate births, and the problems of young mothers (711). In these occupations, Goodenough notes, "the woman psychologist has a decided advantage over the man" (711), presumably because women psychologists would be most effective when dealing with female patients (and would be most likely to be hired for these types of jobs). Goodenough sorts out potential opportunities for her audience, identifying opportunities to fill gaps in the existing research. But these opportunities are mostly feminized ones—areas of research in which women's expertise might be especially valued, and, most likely, areas that would remain stigmatized as "women's work" and downgraded in prestige accordingly.

Although her advice throughout the article depends on gendered assumptions about opportunities available for women, Goodenough concludes by reiterating her view that opportunity and achievement are gender neutral:

> it seems likely that opportunities will not be lacking for women psychologists who take their profession seriously, who are willing to compete with men on an equal basis without demanding special consideration, and who will accept the fact that no amount of faithful work on problems of little importance can compensate for lack of major scientific contribution. Women must cease to rationalize about lack of professional opportunity and demonstrate their competence by actual achievement. Opportunities will expand for those who exert the necessary propulsive force. (712)

Goodenough's editorial provides pragmatic advice for female scientists seeking to contribute to the field of psychology in the postwar years. The declarative mode ("women must," "opportunities will") makes the argument all the more forceful. By exercising foresight, female psychologists could identify the research areas in which they could excel. Women were not advised to seek out positions that would probably be given to men—namely, the high-ranking academic jobs that could lead to prominence within the discipline. Instead, they were advised to carve out niches within traditionally feminine areas of psychological work.

Goodenough paints a gendered picture of postwar opportunities for women scientists, but simultaneously disregards gender as a variable in women's careers. Chosen to lead the NCWP because of her high status in the field, Goodenough was ultimately ambivalent about the organization and wary of advocating any overtly feminist actions. Because her own position in the discipline depended on the approval of prominent men, she supported a noncontroversial approach, one that relied on a discourse of gender neutrality and meritocracy. The opportunities Goodenough identifies were not really gender neutral, but specifically oriented toward conventionally "feminine" abilities and attributes, such as caring for children. Yet Goodenough never abandons the belief that the institutional structure and culture of science *are* gender neutral, and that women's success is simply a matter of "propulsive force" (712).

In their article, Capshew and Laszlo claim that Goodenough simply could not recognize that sex discrimination existed in the field of psychology (165). This unwillingness could account for Goodenough's stance in this editorial. However, it seems that Goodenough's stance was also shaped, at least in part, by the other psychologists with whom she corresponded. Goodenough mentions in the article itself that she consulted with active members of the NCWP, but she does not mention that she also heard from psychologists not associated with that group. Indeed, letters from prominent psychologists in the field may help to explain why Goodenough's editorial seems meant as much to conciliate powerful, conservative members of the discipline as to encourage female psychologists.

Professional correspondence illuminates a hidden factor in the success of a small group of female psychologists, namely, personal patronage. Rossiter writes that, while the climate for women scientists during the years leading up to the war was generally unfavorable, a few women could gain status by aligning themselves with a powerful scientist, often male. In exchange for personal advancements, a woman scientist would need to accept the status quo and not make any waves: "Under such a system of personal patronage, about all the favored women could do was keep up the good work, be loyal to their benefactors, and stay out of 'politics'" (*Women Scientists in America* I 185–86). In other words, the patronage system put women in a defensive position. It allowed for a few women to excel, but did not allow those few women to encourage others to enter the profession on a broader scale or to lobby for institutional or cultural changes.

Before her article was published, Goodenough's male mentor, Lewis Terman, warned her not to take the role of the NCWP too far.[5] On June 27, 1942, Goodenough received the following message from Terman: "I am delighted to hear you were elected president of those female psychologists. I really don't know a thing about their organization [. . .] I believe with Woodworth *et al* [*sic*] that there are probably a number of ways in which this organization can enlarge its contribution during the emergency. *I also agree emphatically with you that it would be unfortunate if the thing were continued as a sort of 'masculine protest' leagued against male psychologists*" (letter to Goodenough, Florence Laura Goodenough papers, emphasis in original). In this letter, Terman implies that the NCWP should stick to carving out wartime opportunities in traditionally feminine areas, as in community-based volunteer work, but warns against any kind of feminist action to improve the status of women in the discipline as a whole. Clearly, according to Terman, it would be inappropriate for Goodenough to advocate any overtly feminist actions.

Goodenough presumably received similar hints from prominent female psychologists, as well. She wrote the following to Martha Crumpton Hardy July 12, 1942:

> My own feeling with regard to the desirability of organizations such as this is quite in accordance with your own [. . .] I feel very strongly that while the whole affair makes relatively little difference one way or the other to those of us who have already found a niche for ourselves, it would be very unfortunate for the younger generation of women psychologists to put themselves in a position where there would be justification for making sex the primary distinction [. . .] So I am entirely in sympathy with your attitude in the matter but I think the best way of handling the situation is through the assumption that the organization is to be only a temporary one but that for the time being it may serve a useful purpose. (Florence Laura Goodenough papers)

As was the case with Terman, we can surmise that Hardy had advised Goodenough against taking the actions of the NCWP too far. This correspondence seems to suggest that some of Goodenough's acquaintances warned her against taking an overly aggressive approach as president of the NCWP. At the very least, these letters would have confirmed Goodenough's own beliefs in the inappropriateness of any overt agitation on the part of female psychologists.

Admittedly, Goodenough actively sought out the advice of trusted colleagues as she prepared her article. Goodenough wrote to Ella Woodward:

> I had a number of chuckles over your letter of May 31st. I think you have hit the nail very precisely on the head when you placed the major responsibility for sex discrimination on the shoulders of the sex allegedly discriminated against. There can be no doubt that many women are demanding a kind of recognition that they have not earned and opportunities for which they have not proved themselves to be qualified. As far as I now there is not, and never has been, any discrimination shown by the publishers of professional journals, for example, against the publication of sound research or significant theory by women authors. Nevertheless, as Dr. Anderson and I will show in a forthcoming article, the amount of publication by women psychologists is very small when compared to that by men. I doubt that this represents a real difference in ability. I think that lack of professional drive and greater concern with matters outside the professional field is largely responsible. (Letter to Woodward, Florence Laura Goodenough papers)

Thus, Goodenough's view of the future for women psychologists was certainly colored by the perspectives of others, who collectively agreed that women simply needed to work harder to publish significant research. Gender neutrality offered an important rhetorical device with which to ward off any critiques or loss of prestige that came with her position as president of the NCWP.

Goodenough's gender-neutral approach may seem untenable and even reprehensible from a feminist perspective, since it does not directly address the institutional causes for women's situation in the discipline. However, Goodenough's article may also be read as a rhetorical balancing act, given the two audiences she addressed. By upholding the gender neutrality and meritocratic ideals of scientific disciplines, Goodenough appeases her prominent readers (and professional supporters). By identifying opportunities for women psychologists, she appeals to members of the NCWP and other women seeking to find a way to contribute to the war effort and to the postwar world.

In fact, Goodenough's advice might be considered "self-defense" for women who, like her, would have to struggle to gain a foothold in a male-dominated discipline. Nonetheless, due in part to her own tendency to rely on advice garnered through personal correspondence, Goodenough was unable to stray from the gender-neutral approach that had thus far stymied the efforts of women scientists to call attention to discrimination against them. The editorial genre tends to reflect expert opinion. In Goodenough's case, her editorial reflected not simply her own opinion but that of other prominent psychologists. Because those experts were invested in the status quo, Goodenough's editorial was unlikely to point to alternative perspectives or possibilities for change.

Georgene Seward: Literature Review

Unlike Bryan and Boring and Goodenough, Georgene Seward approached her article with a sociocultural, rather than individual, perspective on the issue of women in psychology and in American society more generally. In "Sex Roles in Postwar Planning," published in the *Journal of Social Psychology* in 1944, Seward envisions a future in which opportunities could be distributed more equitably for both men and women. Her article most closely resembles a literature review because it draws on a range of sources, including psychological, sociological, medical, biological, historical, and anthropological research. Yet she uses the literature review not just as a forensic genre (one focused primarily on past facts), but also as a deliberative genre, one focused primarily on the future. This genre enabled Seward to draw upon examples from other cultures and other time periods. By looking at a wider range of past histories, Seward was able to envision alternative futures in ways unavailable in a statistical report or even an editorial. In this way, she was better able to see beyond the discourse of gender neutrality that had constrained the other writers.

After a short problem statement, the first section reviews biological differences between men and women, such as size, strength, and reproductive functions. As she cites scientific evidence of gender differences, Seward points out that biology is not static, but subject to change as cultural and social conditions develop. For instance, she notes that while human males are larger and stronger than females, these differences become less important as human societies move further away from hunter-gatherer economies. Through these cultural shifts, she argues, "the ground is prepared for equalizing the status of the two sexes" (164). Similarly, when she discusses reproduction, Seward notes that scientific evidence fails to uphold the cultural assumption that women's performance (both psychological and physical) is impaired by menstruation: "As far as scientific evidence goes, there is no reason why women's reproductive functions should interfere with the performance of social functions. That they often have is largely the outcome of superstition and faulty education" (165). Thus, Seward uses her knowledge of historical and cultural change to suggest that even biologically based sex differences can be attributed, at least in part, to cultural norms, and are therefore open to change.

The second section discusses how cultural norms are similarly open to change. Here, Seward focuses on the cultural construction of masculinity and femininity in various cultures. Drawing on historical and cross-cultural research, she reviews biological, social, and cultural attitudes toward sex roles, surveying, for example, sex roles in ancient Greece, imperial Rome, New Guinean cultures, Nazi Germany, the Soviet Republic, and America. Drawing on an implicit theme of social evolution, she argues that gender roles are not fixed, but subject to change over time, with more advanced societies moving toward greater equality between the sexes: "Since status of women is an index of prevailing social climate, we should anticipate more favorable conditions in the more democratically structured nations as compared with authoritarian forms of government" (179). This argumentative trope stems from Victorian evolutionary anthropological beliefs, (not to mention Social Darwinism), which presume that all societies follow a linear path of development. Thus, the appeal to "social evolution" implies that "primitive" societies represent an "earlier stage" of development than "civilized" ones, an idea already challenged by psychologists and anthropologists like Margaret Mead in the 1930s and 1940s. (Newman 237). However, the logic of this argument persisted, even in feminist theories. Of course, as Schiebinger suggests, history shows that the myth of inevitable forward progress for women does not hold true—"The history of women in science [. . .] has not been characterized by a march of progress but by cycles of advancement and retrenchment" (*Has Feminism* 32). Despite these objections, the appeal to evolution and progress was surely a powerful one, especially for a scientific audience.

In this regard, the passages comparing sex roles in America, Nazi Germany, and the Soviet Union seem most interesting, especially because the article was published in 1944, before the end of World War II and the start of the Cold War. Seward characterizes the enemy, Nazi Germany, as retrogressive, claiming that Nazi women were primarily valued for their ability to breed soldiers (171). There, women were relegated to a more traditional role in which they were valued only for reproduction and child care. In contrast, Seward characterizes Soviet Russia as progressive, since "Women and men enjoy the same education, compensation, rest, leisure, and social insurances" (171) due to extensive economic reforms. There, women's and men's lives were equalized, giving both men and women opportunities to engage in education, work, and relaxation. By placing the United States along a continuum between least progressive and most progressive societies, between Nazi Germany and the Soviet Union, Seward implies that changes in gender roles were not only possible, but desirable. Seward creates an implicit comparison between women in the United States and women in other nations—in this case, women in countries like Italy and Germany—where authoritarian governments prevailed and where retrogressive sex roles were instituted. In doing so, she appeals to notions of American superiority and progress even as she critiques the unequal gender relations within American society.

The third section outlines current research on sex roles, and the fourth section indicates how this research might suggest new roles for men and women after the war. Unlike Goodenough, who advises women to seek out opportunities in areas that are already feminized (and linked to the domestic), Seward urges changes that would allow women and men to pursue careers in any field. She points out that the traditional division of labor in American society is based not on natural ability, but on social conventions:

> That this represents a sex difference in opportunity rather than capacity is suggested by the rapid occupational equation that occurs when external emergency demands it. We have found no scientific evidence in support of a biological sex difference in domestic and cherishing interests. There is no reason to believe that individual differences would not outweigh sex differences if boys were given the same opportunities for developing and expressing interests of this sort. To assign roles on the basis of domesticity seems like an arbitrary and artificial compromise with the *status quo.* (181)

Such a strategy would not help to shift the gendered roles that had traditionally limited women's opportunities. The broader historical and cultural perspective in the literature review genre helps Seward to envision alternatives to women's roles in psychology and in American society more generally.

Seward identifies with feminist approaches to the issue of opportunity in the postwar world: "until feminism is seen as part of the larger movement for human rights it cannot be expected to go very far or to penetrate very deeply" (177). For Seward, it is necessary to change limiting social situations and structures to enable women to engage in public and professional life. But Seward also suggests that patriarchal arrangements place undue burdens on men: "A man's social prestige rests on success. This means preoccupation and specialization in his work to an extent that often excludes the cultivation of family relationships and the sharing of aesthetic experiences. Rarely does the career-woman feel this burden, because she is subtly aware that in the last analysis she will not be judged on the basis of her accomplishment but according to the standards set for her sex. [. . .] Her social measure will be taken in terms of personal attractiveness and the behavior of her children" (175). Because of these narrow gender roles, she argues, women's contributions can arise only in times of crisis: "In a society where the basic problem of production rests upon the men, their security, emotional as well as economic, is threatened by the advent of women in industry, except as temporary emergency replacements" (181). For Seward, a more equal society can be engendered only by altering these stubborn, but not unchangeable, cultural norms.

Finally, Seward argues that the war offered an opportunity to hasten the pace of social evolution: "The present war is accelerating changes in this area as it is causing acceleration in other branches of education. This will mean postwar readjustment of men and women in their social roles and postwar readjustment in our conception of these roles" (175–76). Notably, Seward focuses her attention not just on women, but also on men, suggesting that solutions rested upon increased cooperation between males and females, within both the public and private spheres: "The home in our culture has suffered from being a one-parent institution, and home-making has suffered from the stigma of 'woman's work.' After the security-shattering experiences of the war, our children are all the more in need of stable, well-adjusted, complete homes. This means the restoration of the father as a functional member of the family" (182). Here again, the genre of the literature review helps Seward to assess the current situation, to imagine alternatives, and to argue that the time was ripe for their enactment.

Seward's literature review allowed her to draw conclusions and make arguments that were less available in Bryan and Boring's statistical report or in Goodenough's editorial. By surveying a wide range of cultures and the ways in which gender was configured within them, Seward could envision alternatives to the status quo and call for institutional changes. In contrast, Bryan and Boring's statistical method mapped out the existing state of affairs, and Goodenough's editorial—perhaps because it reflected the opinion of prominent members with an investment in the status quo—similarly served to reinforce a gender-neutral, meritocratic perspective. Unlike a statistical report, which assesses the past, or a primarily epideictic genre such as the editorial, which focuses on a present issue, the literature review genre Seward wrote used past and present research to envision alternative possibilities. In this way, her article is primarily deliberative, and it suggests possibilities for change that are less available in the other two genres.

The Discourse of Gender Neutrality

While Bryan and Boring and Goodenough offer gender-neutral perspectives—seeking to deny or diminish the influence of gender on women's progress—only Seward fully illuminates the prominent role gender plays in shaping social and economic relationships along unequal lines. Ultimately, each of these studies aligns with a different political perspective. Bryan and Boring espouse a conservative approach, suggesting that rapid changes would disrupt the "social equilibrium." Instead, they place responsibility on women themselves to work harder. In contrast, Goodenough seems to espouse a liberal feminist approach. Although Goodenough did not self-identify as a feminist, her arguments "deny that gender differences exist, claiming instead that, for all practical purposes, women think and act in the same way as do men" (Schiebinger, *The Mind Has No Sex?* 275). Such a position assumes that women must assimilate to scientific values and institutional structures in order to make advances, but not that those scientific values and institutions must change to accommodate women. Liberal feminist arguments would imply that women simply needed to work harder and to identify opportune moments in which to advance their interests. Seward's position is the most radical of the three, since it not only exposes the conditions that limit women's opportunities, but argues that institutional changes need to be enacted in order to create more equal opportunities for both men and women. In other words, from a radical feminist perspective, it is important to understand and change the institutional structures that limit opportunities. Rather than simply arguing for women to work harder to achieve status, such a perspective suggests that institutional and cultural changes need to be made in order for all individuals to have equal opportunities for advancement.

After the war, it became apparent that arguments along the lines of the ones Bryan and Boring and Goodenough made carried the day. While women were disproportionately represented in the lower ranks of psychologists, and were generally unable to address the systemic factors that contributed to their underrepresentation, a few women were able to take advantage of the wartime situation to make professional gains. Aside from Bryan, Goodenough, and Seward, several other women contributed to wartime psychology, often in the realms of testing and recruitment. Eugenia Hanfmann, originally from Germany, was hired to help conduct rigorous psychological testing for recruits of the Office of Strategic Services (Hanfmann 144). The United States Department of Agriculture hired a number of psychologists to work in the Division of Program Surveys to research American attitudes to a range of wartime problems (Hyman 3). Countless others contributed to psychological research and outreach in their own communities.

Overall, however, the NCWP was not able to enact changes in the discipline of psychology itself. Capshew and Laszlo claim that the high degree of professionalization among women psychologists may have actually limited their abilities to advocate against discrimination: "Thoroughly socialized by an elite scientific and professional ideology, the majority of women psychologists found it more palatable to attribute their low status to low achievement than to any conscious discrimination" (176). Indeed, gender-neutral discourses seemed to be so powerful that they overwhelmed the objective evidence gathered by Bryan and Boring, which suggested that women were systematically channeled into less prestigious areas of the discipline. Instead, some women psychologists agreed with Goodenough that their lack of advancement was due to their own lack of "propulsive force."

Further, assumptions about the culture of science persuaded women to continue as they had in the past. The longstanding belief among scientists was that the scientific method guaranteed future scientific progress, that publishing scientific articles would lead to promotion, and that this method was meritocratic and gender neutral. To suggest a gender bias with relation to opportunities in science would require female psychologists to question or reject the cultural beliefs about science that they had internalized.

Another explanation of female psychologists' failure to enact significant changes in the discipline stems from a disagreement about the nature of the opportunity the war offered—particularly, about whether it was appropriate to foreground women's concerns (let alone specifically *feminist* concerns) during a time of crisis. Although most agreed that the war presented *some* form of opportunity for women, female psychologists disagreed about the extent to which it presented the right moment to argue specifically for women's rights. In a retrospective account, Bryan recalls that the NCWP was founded with a specific focus on employment, not radical action: "Because at the time of the founding of NCWP most of the prospective members had full-time jobs, to some *the time did not seem propitious* for militant action aimed at securing higher status, better-paying positions for women only" (183, emphasis added). Similarly, although the ECP subcommittee recognized that the issue of women's roles in psychology would persist, it did not seem appropriate to address these problems during the war. In 1943, ECP member Ruth Tolman wrote that postwar problems "will demand [. . .] attack on a wide front, with all the resources the profession can command. It seemed hardly feasible, therefore, for the Subcommittee *at this time* [. . .] to undertake the planning for such future responsibilities" (297, emphasis added). The committee felt that the war required women to address pressing wartime issues, rather than to agitate prematurely for postwar changes.

As the war came to a close, the exigence surrounding these initial efforts seemed to diminish even more. One reason for this, ironically, may have been the success of women psychologists in finding wartime work. Tolman wrote that by August 1943, the subcommittee of the ECP on the services of women psychologists voted to disband because "the needs of women psychologists were now being adequately and competently handled by existing organizations" (297).[6] While the NCWP did not disband after the war, its focus did shift considerably. Membership in NCWP had begun to drop toward the end of the war, perhaps because the exigence attached to the war itself diminished as victory seemed more and more certain (Cautley 4). However, several psychologists from Canada and Europe expressed interest in joining the organization. At a 1946 board meeting of the NCWP, Louise Ames Bates proposed that the name of the organization be changed to the International Council of Women Psychologists, and the motion carried (Hogan and Sexton 628). Accordingly, the organization changed its focus to furthering international and intercultural relations in psychology (Russo and O'Connell 46).

Following the war, the ICWP continued its attempts to foreground the activities of women psychologists, in part by sponsoring special sessions at the APA's annual meetings, but many members feared that overt questioning of women's unequal status in psychology would be poorly received (Walsh 20–21). The ICWP decided that gaining recognition as an official division of the APA would best help women to gain status, but the APA refused them in 1948, stating that the APA would not admit a division composed of just one sex (21). By 1959, the organization voted to drop the word *Women* from its title and to admit male members, in part in an appeal to gain official affiliation with the APA (which was nonetheless rejected because the international focus of the group did not suit the Association's mission) (Hogan and Sexton 628; Walsh 22). It was not until 1973 that the APA formed a division on the psychology of women, an organization that self-identifies as feminist in its purposes but explicitly welcomes both male and female researchers (Walsh 23).

The supposed "failure" to enact feminist reforms during World War II is not due simply to an inability to seize an advantageous moment, but has more to do with the complexity of the situation and with the deeply entrenched beliefs about the gender neutrality of science. For a fledgling field interested in branding itself as "scientific," the discourse of meritocracy served as a powerful trope, one that deflected attention away from the inequalities within scientific institutions. Only by critiquing and rearticulating *scientific ideals* (meritocracy, gender neutrality, and so on), genres, and institutional structures could female psychologists fully address the unequal opportunities women faced in psychology. Yet questioning these ideals and practices might have been unpopular, especially during a time of war, when victory seemed to be the most important goal.

2. *Women Anthropologists Study Japanese Internment*

> I have been terribly disappointed with the results. There is something lacking in my reports which constantly frustrates me. Whenever I proofread my reports before sending them to you I feel like an outsider peering at the evacuees from a telescope. I had hoped in the beginning that being a member of the "in-group" by necessity, I would be able to make an "ideal" study of it but the results have been far from satisfactory. Perhaps the detailed report will help to portray the evacuees as a complete picture rather than bits of humanity thrown indiscriminately in space as I have been picturing them.
>
> —Tamie Tsuchiyama to Dorothy Thomas, 3 January 1944

> I'm really afraid that I have been spending too much time snagging material from the administration and making copious notes. I should be writing more but I got such a great deal to learn these first few months that getting material seems more important than writing matters on which I am incompletely informed.
>
> —Rosalie Hankey to Dorothy Thomas, 19 August 1943

Shortly after Pearl Harbor, in the early months of 1942, the United States Department of Justice began preparations to evacuate "alien enemies" of Japanese descent from the West Coast. On February 19, 1942, President Roosevelt issued Executive Order 9066, which gave the army the authority to identify crucial areas from which citizens or aliens would be removed. By March 16, 1942, the government had designated the entire West Coast a restricted area for all Japanese-Americans. The next day, the War Relocation Authority (WRA) was established to oversee ten isolated camps, where Japanese-Americans would live, many for the duration of the war. Soon, 110,000 people were removed from the West Coast, first to a series of temporary "assembly centers," and later to one of ten "relocation centers" (Hayashi 1). Sociologist Dorothy Swaine Thomas and a colleague, Charles Aiken, seized this opportunity to study the sociological effects of internment, launching the Japanese Evacuation and Resettlement Study (JERS) (Thomas, "Experiences" 667). Soon afterward, as the evacuation plans expanded into a large-scale, forced evacuation of all residents of Japanese ancestry, Aiken left for another wartime project, which left Thomas as the sole director of the project. Funded mainly through private sources, the JERS lasted from February 1942 to December 1945 (Thomas and Nishimoto vi).

Thomas enlisted a cohort of "informants" to conduct research on life at the internment camps, including Tamie Tsuchiyama and Rosalie Hankey, both graduate students in anthropology at the University of California. Their writing has to date received little attention from scholars in the history of science, yet it can tell us much about the rhetorical and personal effects of anthropological genres. Anthropologist Ruth Behar has written that "When a woman sits down to write, all eyes are on her. The woman who is turning others into the object of her gaze is herself already an object of the gaze" ("Introduction" 2). Both Tsuchiyama and Hankey struggled to efface their gendered subjectivities to produce appropriately "distanced," objective accounts of their research.

In this chapter, I examine how the genres at work on the JERS study called into play different speaking (or writing) subjects. Feminist researchers have argued that identities do not simply exist prior to any rhetorical situation, but are constituted and reconstituted through discourse events. Barbara Biesecker argues that, rather than asking "Who is speaking?" feminist scholars should consider a different question: "what play of forces made it possible for a particular speaking subject to emerge?" (148). The JERS study provides a compelling illustration of how identities emerge through scientific discourses and genres because Japanese Internment helped to constitute new subjectivities: Americans of Japanese descent were called to identify themselves as Japanese first and foremost—even those who had grown up in America, speaking English and identifying primarily with mainstream American values. Similarly, white researchers in internment camps came to understand themselves as Caucasians, as outsiders and potential spies.

Yet, the field report genre required Hankey and Tsuchiyama to subordinate these multiple, and often conflicting, subjectivities in favor of the stable identity of an "objective field researcher." As the various types of documents involved in the study—transcripts, field notes, field reports, and so on—moved further away from the field (culminating in Thomas's comprehensive books, *The Salvage* and *The Spoilage*), their language became increasingly abstracted from the internment camp. Thomas required her informants to write using rhetorical features that encouraged detachment in the guise of objectivity. For this reason, the shifting identities that emerged during fieldwork were subordinated, and the writers were encouraged to employ strategies of "temporal distance"—rhetorical techniques that distanced the observer from the time and place of their observations. Their personal reflections and opinions appear only sporadically, through occasional attempts at "ethnographic presence," the set of linguistic and rhetorical devices ethnographers use to establish a lively, realistic, and detailed account of the field. Both Hankey and Tsuchiyama struggled with the objectivity Thomas mandated and ultimately failed. Not only did the discourse of objectivity take a personal toll on Hankey's and Tsuchiyama's careers; it also suppressed critiques of the ethics of internment that these researchers might have made. Neither researcher was permitted to use the field report to offer personal reflections on their experiences in the field, even though these reflections might have offered valuable insights into the sociological effects of internment. The JERS study tells us much about how young scholars learn to assimilate the dominant genres in a research area, and how those genres function to suppress the very kind of "situated knowledges" feminist scholars would argue are necessary for more responsible kinds of scientific research and writing (Haraway, *Simians* 188).[1]

The Culture of the JERS Study

Between 1942 and 1945, the War Relocation Authority (WRA) established ten concentration camps, ranging in size from approximately 7,000 to 19,000 people (Ng 38). Evacuees lived in wooden barracks, organized into blocks where 250 to 300 people lived in cramped quarters (Hayashi 1). At each camp, a staff of approximately one hundred oversaw administration. The administration appointed bilingual block members to disseminate goods, services, and information, and formed an elected community council to legislate on management issues (1). Any in the camp would have been automatically identified as members of the administration, whether they worked for the administration itself, the JERS, or for another project, the Bureau of Sociological Researchers (BSR). This second group of researchers was charged specifically with facilitating relocation by pressing evacuees to agree to a Loyalty Registration, a mandatory survey that asked individuals to pledge allegiance to the United States and a willingness to serve in the American armed forces (Ng 57).

The administration sought to reproduce some semblance of mainstream American community life in the camps, with the ultimate goal of identifying and relocating those evacuees who proved their loyalty to the United States government. Evacuees worked for low wages in the camp itself (i.e., as butchers, firemen, cooks, nurses aides, etc.), on farms, or on government projects (i.e., making camouflage nets for the military). Children attended school. Evacuees played baseball, participated in golf tournaments, attended dances, and published newsletters. In short, administration tried to create new communities that would teach American values. Clearly, though, these "communities" betrayed key American rights such as equality and justice. Entries and departures were strictly controlled by Military Police (Hayashi 92). Evacuees lived in cramped quarters, often in extreme heat, with poor food, little pay, and even less freedom.

Nonetheless, beyond this "official" organization, evacuees reproduced prewar community organizations that operated under the radar of the Euro-American administrators. While the WRA excluded the first generation, or Issei, from official political participation in the camps, prominent members of the Issei community organized unofficial groups that would lobby the WRA and organize evacuees around common concerns (e.g., fairness of wages or quality of living conditions). These groups were dominated by men, so it was difficult for Tsuchiyama and Hankey to infiltrate them. Researchers such as Tsuchiyama and Hankey were to report on all aspects of camp life, but Tsuchiyama, especially, found this unofficial political activity particularly compelling.

Of course, Thomas, Tsuchiyama, and Hankey were positioned differently with relation to the research, due to factors such as race, gender, space, time, and status, and these factors impinged on the genres they wrote and the subjectivities called into play through their research on the JERS study. Thomas, an accomplished sociologist, had trained as an undergraduate at Barnard College and earned her Ph.D. in 1924 from the London School of Economics (Hirabayashi 61). She held positions at the Federal Reserve Bank, the Teachers College at Columbia University, and Yale University's Institute of Human Relations before moving to the University of California at Berkeley in 1941. As a senior researcher and head of the JERS project, Thomas did not conduct field work of her own. Instead, she received field notes and reports from her staff members stationed in the camps. She remained sheltered at the university, where she had the assistance of an administrative staff, the comfort of her own home and office, and the authority to determine how the field workers would conduct their research and write their reports (Hirabayashi 168). Meanwhile, Tsuchiyama and Hankey were each stationed in internment camps, where they lived in barracks and ate in mess halls.

Thomas organized the JERS study to maximize the amount of data collected: she dubbed it the "vacuum cleaner approach." Informants gathered documents of all kinds, including administrative instructions, camp newspapers, minutes of meetings, notes from interviews or meetings with evacuees, and letters of complaint generated by the evacuees (Thomas, "Experiences" 668). Most of the researchers also kept "undirected journals," where they wrote about day-to-day events at the camp, took down the information they had gathered from the evacuees (such as verbatim statements), and recorded their own experiences and attitudes (668). Given the different approaches of each researcher in gathering and sorting information, Thomas writes, "Each brought to his journal something of the standpoint of his own discipline and his own biases, in the very process of selecting events, words, and acts to record" (668). Presumably, it was Thomas's job to see through those biases to develop an authoritative account of life in the internment camps. In addition, each informant prepared comprehensive reports on topics such as the segregation of "loyal" and "disloyal" evacuees, or the evacuees' reactions to particular events. Finally, the researchers wrote letters to Thomas, updating her on their progress and on conditions at the camps. Sometimes, these letters offered more personal reflections than did the field notes or reports (as was the case with Tsuchiyama); in other cases, the letters seemed more like field notes, with section headings and reporting mixed in with more personal elements.[2]

The "vacuum" metaphor is indeed apt; the general flow of information was "up the chain of command." Field reporters sent large quantities of "undigested" information to Thomas, who collated, synthesized, and analyzed what they collected. Based on this large body of field research, Thomas wrote two books, *The Spoilage* (1946, with Richard Nishimoto) and *The Salvage* (1952, with Charles Kikuchi and James Sakoda as coauthors). *The Spoilage* refers to the 18,000 evacuees who refused to declare their loyalty to the United States, while *The Salvage* refers to the remainder, who declared loyalty and left the camps. Despite the seemingly critical titles, Thomas and her coauthors sought to provide a neutral account of internment. The books did not overtly criticize the internment policy. Instead, they assumed that those Japanese-Americans who assimilated to American culture were the norm, and that those who failed to do so were in some way deviant.

Both of the young researchers found it difficult to get started on their reports. Hankey wrote to Thomas that she was overwhelmed with information: "I'm accumulating material much faster than I can write it up. I still have about three folders of notes from my first stay here [. . .] I've worked every day including all Sundays" (Hankey to Thomas, 19 October 1943). Tsuchiyama often began her letters to Thomas with apologies for not being more productive and promises that her next report would soon be finished. For example, on November 19, 1943, she wrote to Thomas from Cleveland, where she had gone to take a break from camp life: "I am gradually pulling myself together to make a stab at the work I have set out to do in the next few months. Aside from going through my files and attempting to bring my notes up to date I have accomplished very little since coming out. At present I am working on an outline for an 'ideal' study of Poston to give me an idea what gaps will have to be filled when I return to Poston" (Tsuchiyama to Thomas, 19 November 1943). Both researchers ended up producing significant work on the project, but their letters suggest that writing these reports was a struggle. In what follows, I consider how the material conditions of camp life and the demands of the JERS study placed Tsuchiyama and Hankey in conflicted positions, positions that complicated the seemingly straightforward task of writing field reports.

Tamie Tsuchiyama: Competing Subjectivities

Tsuchiyama was a young graduate student working on a Ph.D. at the University of California, Berkeley, and the only Japanese-American woman hired full-time for the JERS project. She was a second-generation American; her parents had emigrated from Japan and settled in Hawaii, where they ran a small vegetable farm. Her father had been a newspaper editor in Japan, while her mother came from a family of samurai lineage (Hayashi 56). In 1936, she left her home state of Hawaii to study anthropology at the University of California, Los Angeles as a bachelor's student, and later at Berkeley, where she studied with renowned anthropologists A. L. Kroeber, Robert Lowie, and Paul Radin (Ng 150). In April, 1942, Lowie connected her to Thomas (Hirabayashi 24). In May, she relocated to the Santa Anita Assembly Center, a temporary camp in California, and began collecting data for Thomas. In August, she moved to the Poston camp in Arizona, where she remained until October of 1943. Poston was the second-largest camp, with a population of 17,814 evacuees at its peak (Ng 38).

Tsuchiyama might have expected to have an advantage when it came to field research in the internment camps, since she was herself

Japanese-American and could speak and write Japanese with some fluency. However, as a second-generation Japanese-American, or *Nisei,* she was distanced from the first-generation, or *Issei,* who also populated the camps. Like many second-generation immigrants, she grew up speaking both Japanese and English and was encouraged to become Americanized.[3] Further, Tsuchiyama grew up in Hawaii, where a different set of racial and ethnic relations obtained; historian Wendy Ng argues that the Hawaiian Japanese may have felt more integrated into life in Hawaii than their mainland counterparts felt with regard to the continental United States (11). Further, Tsuchiyama was different from many of the evacuees by virtue of her education and social background. In addition to Japanese, she was competent in German, Spanish, Italian, Chinese, and several Polynesian dialects; she was engaged to a Euro-American graduate student; and she was a member of the American Civil Liberties Union (Hayashi 56). Thus, Tsuchiyama's race, family background, ethnicity, and class would come to the fore in different ways during her experience at Poston, depending on contextual factors.

Her situation in the camps also accentuated her identity as a woman. Tsuchiyama found it difficult to infiltrate the male-dominated political organizations that emerged at Poston. Some of these were implemented by the WRA to encourage "self-government"; not surprisingly, the WRA appointed only American citizens to the community councils they designed (Hayashi 111–12). Other organizations were formed by evacuees themselves, and Tsuchiyama found that these were often dominated by the *Issei.* Tsuchiyama forged a friendship with Richard Nishimoto, a powerful member of the community, in order to find out about these more secretive camp activities. Her close involvement with Nishimoto led some to question whether the two were having an affair. Living and working in such close quarters meant that gender became salient in ways that might not have mattered under different circumstances.

To make matters worse, residents of the internment camps were aware that they were being monitored by the FBI and were generally suspicious of anyone who might be considered an informer, or *Inu* (literally, dog) (Hayashi 125). If others found out about Tsuchiyama's research, she could easily be branded an *Inu* and be subjected to intimidation or worse. As a *Nisei* female researcher, Tsuchiyama's position was a particularly vulnerable one, given the tense and sometimes hostile camp situation.

Tsuchiyama seemed frustrated with her inability to provide a holistic, detailed, and "human" account of the evacuees with whom she had been living. Yet in contrast to Tsuchiyama's own self-deprecating account, Thomas indicated that most of Tsuchiyama's reports were "outstanding" ones that took "a given subject and explored it thoroughly," while providing "background and a consistent follow-through" (Thomas to Tsuchiyama, 28 July 1944). In spite of these reassurances, Tsuchiyama became increasingly dissatisfied with her work at Poston and with the JERS study itself. In 1944, after a series of skirmishes with Thomas, she left the study for good. Although parts of her reports appear in *The Spoilage,* the book Thomas published after the war, Tsuchiyama herself never published the results of her research. Instead, she abandoned the topic, focused her dissertation research on Native American mythology, and ultimately left the field of anthropology altogether.

What might account for Tsuchiyama's own dissatisfaction with her reports? The genre of the anthropological report Tsuchiyama adopted required her to employ "temporal distance" in her writing, a stance that alienated her from the events she experienced at Poston. By temporal distance, I mean the linguistic and rhetorical strategies ethnographers use to shape a coherent, objective account of the multiple, sometimes incommensurable, daily events in which they were submersed. Anthropologist Johannes Fabian remarks that "Generally speaking, anthropology appears to have been a field of knowledge whose discourse requires that its object [. . .] be removed from its subject not only in space but in time" (198). Indeed, the credibility of an ethnographic account depends on the fact that it has been extracted from the firsthand observations and events upon which it reports (Thornton 298). In the act of writing, the ethnographer is situated in a different time (and often place) from the event itself. According to George E. Marcus, this is how the ethnographic genre works; ethnographic writing is "temporally and intellectually separate" from fieldwork (171), "deskwork" as opposed to "fieldwork" (171). The temporal and spatial distance from the field purportedly assures that the ethnographer's account will be more "objective"—time and space seem necessary to sort through observations and account for them from a scientific viewpoint. However, temporal distance limits the ethnographer's ability to situate herself with relation to the people and events she is describing or to reflect on the ways in which her own subjective experiences might illuminate her interpretations. Perhaps it was this rhetorical feature, this sense of temporal distance, that made Tsuchiyama so unhappy with her reports and the way they seemed to depict camp life "through a telescope."

Temporal distance emerges in Tsuchiyama's reports through three primary rhetorical features: an impersonal form of reporting, the use of indirection to express personal opinions, and chronological organization. First, Tsuchiyama relies on impersonal language in much of her reports. For example, her report titled *The Beating of S_____ K_____* chronicles official and unofficial accounts of an incident that occurred in February, 1943, where eight evacuees attacked K_____, an interned Japanese-American man who had been cooperating with U.S. army officials seeking new recruits (*Beating* 2).[4] In this report, Tsuchiyama provides an account of the event as "reconstructed from data supplied by my Camp II informants" (2). These informants are not named in the report, so much of it is written in the passive or impersonal voice. Take the following passage, for instance:

> The posting of guards *is reported to have occurred* as a result of a threatening letter which K_____ received about this time. His popularity in Camp II has never been very great but resentment against him increased noticeably when the War Department announced plans for recruiting a Nisei unit on January 27. At that time *a number of people recalled that* K_____ had participated in the JACL convention in Salt Lake City in November when the resolution concerning selective draft had been framed and *they also remembered that* he had been rather active passing out petitions to Nisei to volunteer for the U.S. army. (2, emphasis added)

By downplaying the sources of her information, Tsuchiyama deflects attention away from the identity of the informants and how reliable their versions of the events might be. By the time she describes the actual beating of K_____, Tsuchiyama has slipped into a third-person, "omniscient observer" narrative form: "M_____ entered the apartment first and as soon as K_____ perceived him he called out: 'You are M_____, aren't you?' This completely unnerved M_____ giving K_____ an opportunity to leap out of bed and pin him down. At this moment T_____ came in and proceeded to whack K_____ on the head and shoulders. When the victim lifted one hand to ward off the blows two of his fingers were smashed" (4). Without the metadiscursive markers indicating where this information came from, this section of the report creates a more immediate account of the events. The reader becomes drawn into the intrigue of the beating, and the narrative voice renders Tsuchiyama herself temporally distant, even invisible—she does not indicate whether she was herself present during the events described, or even whether any of her informants had actually witnessed those events.

In addition to the impersonal tone, Tsuchiyama also uses indirection to deflect attention away from her own interpretations of events. For example, in her report entitled *Chronological Account of the Poston Strike,* she describes how a Mr. Evans—the acting project director—attempted to address the strikers over a P.A. system, but was disrupted when someone unplugged the device:

> There was a great commotion after the translation was completed. Curiously enough, the first to razz Evans was a young Nisei girl about 18 years of age. Evans realized that what he had said would not be heeded. As he proceeded to leave the police station he sighted a group of ten to fifteen block managers at one side of the building and began to talk to them. He received a cold reception from them. *It was one of the most tragic sights to many people* since up to then as head of the block manager system he had been in intimate contact with them. *It was quite apparent that* the block managers were uneasy and did not wish to appear friendly with any member of the administration. (*Chronological Account* 5)

In the passage above, Tsuchiyama relies on impersonal constructs ("It was one of the most tragic sights," "It was quite apparent") to carry the more subjective aspects of her observations, rather than putting them in her own voice. It is unclear whether Tsuchiyama herself identified with those who considered this a "tragic sight," or whether she identified with those who gave Evans a "cold reception."

For the most part, Tsuchiyama veils most of the statements that could be taken as illustrative of her own opinion, especially if that opinion could be construed as critical of internment or of the administration. In her report entitled *Notes on Selective Service Registration,* Tsuchiyama describes a meeting that took place on November 17, 1942, where some American sergeants were attempting to persuade Japanese-Americans to enlist in the army. Tsuchiyama writes: "They explained that their system was to send a Kibei and a Jun Nisei together into the field, the Kibei translating Japanese into English, and the Nisei vice versa. *The underlying insinuation as interpreted by the audience was* that the U.S. army could not trust the Kibei but that they were indispensable because of their mastery of Japanese" (*Notes on Selective Service* 1). Here, Tsuchiyama does not mention whether she was part of the audience at this meeting, or whether this information was received from her own informants. It is hard to tell whether Tsuchiyama interpreted the sergeant's appeals in the way she describes. Certainly, stating such an opinion openly would have placed Tsuchiyama in a delicate position, given her own careful attempts to portray herself in a pro-American light. For instance, on the second page of the report she writes, "The atmosphere was definitely antagonistic toward the poor Japanese sergeants" (*Notes on Selective Service* 2). The word "poor" suggests that Tsuchiyama sympathized with the army officers, rather than the crowd. Only later in the report does it become apparent that Tsuchiyama was indeed present at the meeting: "X who was standing beside me remarked that the sergeants did not have sufficient command of the language to be able to translate or to question

Japanese prisoners" (3). By nearly effacing her own presence at the meeting, Tsuchiyama also effaces her own attitudes, which become apparent only through subtle cues in her word choices.

One explanation for Tsuchiyama's use of passive and impersonal constructs may have to do with the nature of the research she conducted. Many of Tsuchiyama's reports focus on the political organizations and intrigues that cropped up among the evacuees. Tsuchiyama relied heavily on Nishimoto, (deemed "X" in the reports), a powerful man and her chief informant. However, Tsuchiyama's tendency toward indirection may also be a reflection of her own tenuous position as a Japanese-American researcher. Tsuchiyama would need to demonstrate her loyalty to the United States throughout her writing, or she might herself be suspected of disloyalty.

Using indirectness and implication as a kind of "rhetorical ventriloquism," enables Tsuchiyama to express points of view that might not otherwise seem appropriate for a Japanese-American. For example, later on in the same report, she recounts a scene in which a group of kitchen staff workers had resigned in solidarity with the strike:

> Afterwards I heard that none of the staff showed up at breakfast next morning so the Caucasians were compelled to prepare their own breakfast. When the delegates arrived to negotiate with the administration Thursday morning the spokesman for the administration reminded them that after all they were not cutting off the evacuees' food supply and they could at least reciprocate by working in the personnel kitchen. The delegates carefully explained that the kitchen staff had not quit because they had been ordered to but for personal reasons—that they resented the superior attitude of the Caucasians toward them. (*Chronological Account* 12)

Normally, the Caucasian staff members would have been served by evacuees working in the mess halls. Tsuchiyama is careful not to sympathize with the strikers overtly, but one could interpret this passage as a sort of indirect jab at the Caucasian administrative staff at the camp, who had been forced to make their own breakfasts because of their "superior attitudes." Yet, Tsuchiyama's careful phrasing also makes it hard to pin this down with any certainty.

Tsuchiyama also creates temporal distance by organizing her reports chronologically by date. For instance, her "Chronological Account of the Poston Strike" report includes no introduction, but begins in medias res with a description of events on Sunday, November 15, 1942. Tsuchiyama does not situate herself with relation to the events on this date; instead, she begins with a factual report of the incidents leading up to the strike:

"G_____ F_____ and I_____ U_____, both of block 28, were arrested Sunday afternoon about 1:30 and placed in the Poston City Jail by Ernest Miller, Chief of Internal Security, presumably in connection with the beating of K_____ M_____, so-called F.B.I. informer, the evening before. At that time friends of the two Kibei boys recalled that F_____ had had a big scrap with Horris James, Press and Intelligence Officer, toward the end of August" (*Chronological Account* 1). The report continues with notes on the events that occurred each of the following days. The dates themselves provide section headings, and the length of each entry ranges from a few paragraphs to over ten pages (for events that occurred on Wednesday, November 18, 1942).

This chronological organization would have been imposed by Tsuchiyama after the fact on disparate bits of information, such as her field notes (an example of which is shown in Figure 2.1), copies of documents she gathered, and so on. Because Tsuchiyama was able to retrieve a large amount of data from different informants on different days, she would have had to sort out various accounts, documents (such as newspaper reports or meetings from minutes), and transcripts to arrange them chronologically. Yet the chronological organization itself erases the hand of the writer; it appears to be neutral and automatic, not an act of authorial interpretation.

Further, the chronological organization subordinates Tsuchiyama's own subjective experiences of life at Poston. Philosopher Henri Bergson writes that humans experience time as "pure duration," or the "the form which the succession of our conscious states assumes when our ego lets itself *live,* when it refrains from separating its present state from its former states" (100). In other words, we do not often arrange our experiences along a strict chronology, by hour or day; consider the difficulty most of us have in pinpointing the precise dates and times of even relatively important events in one's life. In practice, the distinctions between past/present/future melt together in duration, which includes both the past and present as part of an "organic whole" (Bergson 100). Nonetheless, rational human thought tends to spatialize time, so that duration is converted into a series of discrete moments marked by linear time (seconds, minutes, hours, days, etc.), "a continuous line or a chain, the parts of which touch without penetrating one another" (101). A chronology is one type of line or chain used to rationalize and spatialize time, but it is not a neutral or automatic one. As Kenneth Gergen writes, "Linear temporal ordering is, after all, a convention that employs an internally coherent system of signs; its features are not required by the world as it is" (251). That is, chronological time offers just one possible way of or-

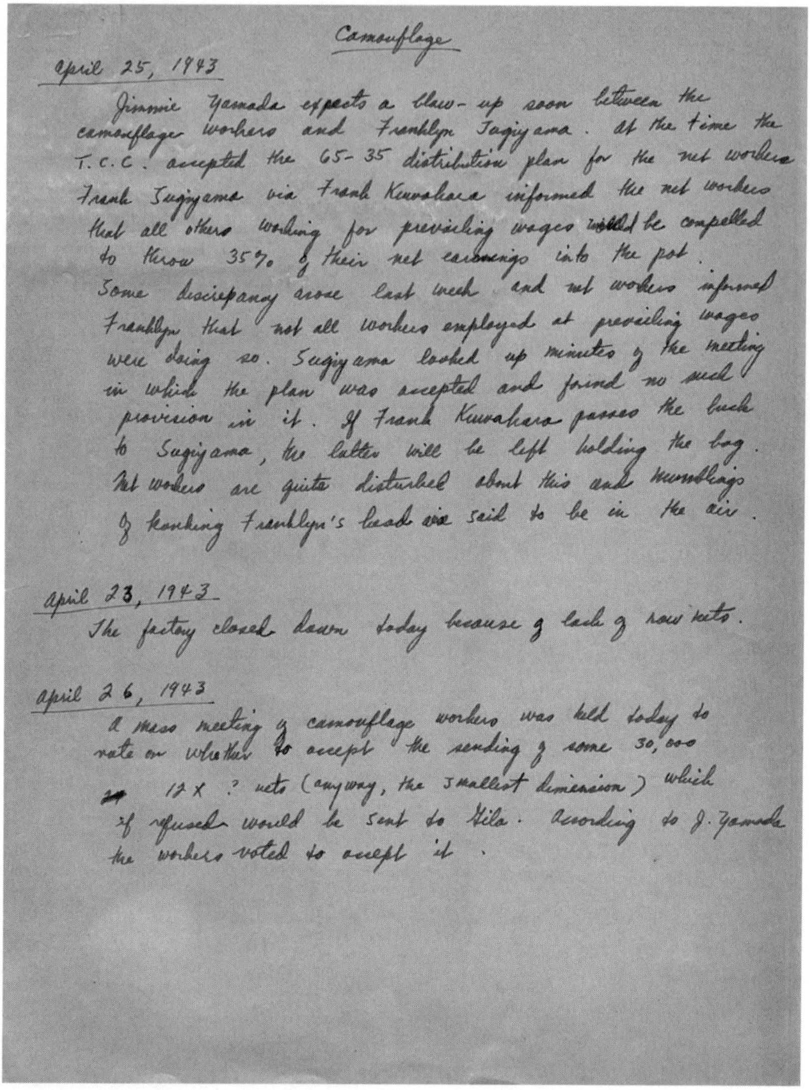

Camouflage

April 25, 1943

Jimmie Yamada expects a blow-up soon between the camouflage workers and Franklyn Sugiyama. At the time the T.C.C. accepted the 65-35 distribution plan for the net workers Frank Sugiyama via Frank Kuwahara informed the net workers that all others working for prevailing wages would be compelled to throw 35% of their net earnings into the pot. Some discrepancy arose last week and net workers informed Franklyn that not all workers employed at prevailing wages were doing so. Sugiyama looked up minutes of the meeting in which the plan was accepted and found no such provision in it. If Frank Kuwahara passes the buck to Sugiyama, the latter will be left holding the bag. Net workers are quite disturbed about this and mumblings of conking Franklyn's head are said to be in the air.

April 23, 1943

The factory closed down today because of lack of raw nets.

April 26, 1943

A mass meeting of camouflage workers was held today to vote on whether to accept the sending of some 30,000 12 x ? nets (anyway, the smallest dimension) which if refused would be sent to Gila. According to J. Yamada the workers voted to accept it.

Figure 2.1. A page from Tamie Tsuchiyama's field notes; image courtesy of the Bancroft Library, University of California, Berkeley.

ganizing experience, one that has become conventional in modern life, especially because it has been institutionalized in technologies such as calendars, clocks, and schedules.

Although chronologies are useful for rational thought, planning, and organization of events, they ultimately diminish or subordinate the more fluid, subjective sense of time as duration. The chronological organization Tsuchiyama used in her reports is therefore not a neutral or automatic choice, but a rhetorical use of quantitative time, or *chronos,* that serves to organize a series of disparate events. The chronological organization both enacts and requires temporal distance from the events themselves.

Overall, Tsuchiyama's reports subordinate her own subjective reflections and participation within the field. This neglect of the personal is especially striking given that Tsuchiyama would have been interned had it not been for her status as a researcher. While her own views and experiences could easily be considered relevant to the JERS study, Tsuchiyama did not feel comfortable assuming an auto-ethnographic approach.

Indeed, when Tsuchiyama does use the more personal, first-person voice in her writing, it is usually in a metadiscursive capacity, as a signal to help the reader interpret the text. For example, in her *History of the Central Executive Committee,* Tsuchiyama writes: "In my paper on the aftermath of the strike I mentioned that M______, Issei representative from block 45 approached X on Wednesday, Dec.2, and informed him that there was much dissatisfaction [. . .] I also reported that this group was successful in intimidating a large number of the delegates" (*History* 2–3). Later, Tsuchiyama calls attention to her own translation of materials she had obtained in Japanese: "I have tried to keep the translation as literal as possible" (17). This metadiscursive "I" situates Tsuchiyama not within the context of the field research, but outside of it, as someone who is reading and arranging her observations after the fact. In other words, the "I" does not increase ethnographic presence in Tsuchiyama's writing. Instead, it increases temporal distance by providing organizational markers for readers rather than subjective interpretation of events.

Occasionally, Tsuchiyama does include her own observations, written in the first person, to call attention to the ways in which her own gendered or racialized identity emerged in her observations and experiences at Poston. Tsuchiyama hints at her own discomfort as a researcher who, if branded an *Inu,* could be subject to intimidation or worse. For example, in her report on a strike that occurred in the Poston camp in November 1942, Tsuchiyama describes the air of suspicion and hostility among the evacuees. She reports that on Friday, November 20, she noticed that warnings against informers had increased, including a number of images posted around the camp, such as "a cartoon of a dog as well as an authentic bone bearing the message 'This is for dogs' on the police station wall" (*Chronological Account* 22). She also mentions that she received a warning while sitting around the camp fires among the protesters: "A middle-aged woman sitting next to me remarked: 'They say there is a female dog in our block. I wonder who she is.' (I thought: 'She needn't be so damned pointed about it' but later discovered that the female dog she referred to was not I but another girl who worked in the Red Cross office)" (23). Given the tense events during the strike at Poston, Tsuchiyama probably felt more anxiety than she admits in this passage.

When Tsuchiyama does include personal reflections on events, they mostly serve to downplay her identification with the more traditional Issei and to assert her allegiance to the United States. For example, in her report, *Notes on Selective Service Registration,* Tsuchiyama recounts a speech she heard by a military recruiter on February 17, 1943. Because this passage is one of the only ones in which Tsuchiyama brings her identity to the fore in this manner, I will quote from it at length:

> The ceremony opened with the singing of the national anthem with the army team saluting a huge American flag draped in the background. I agree with A_____ K_____ that the American flag never looked more beautiful to me than it did in that blistering Arizona sun [. . .] Even the national anthem which in pre-evacuation days had infallibly given me indigestion seemed to have a new meaning and I felt like shouting defiantly to the people about me when we came to the section: "And the star spangled banner, long may it wave/ Over the land of the free, and the home of the brave." That sense of isolation—that feeling of being cut off from the rest of the world—which has been gnawing within me for the past few months completely engulfed me at the moment and I experienced an insane desire to escape to "America" as rapidly as I could. I am recording my reactions, which are by no means unique but shared by a number of Nisei, because I consider them significant in "placing" the position of pro-American Nisei caught in the relocation centers today. The Nisei in Poston have been so subjected to Issei domination that I doubt if very many would be totally taken by surprise to see a Japanese flag flying over the rooftops one of these mornings. (*Notes on Selective Service* 13–14)

Here, Tsuchiyama blends her account of an experience in the field with her later analysis of that experience (although she does feel compelled to justify her personal reflection as representative of others' opinions). Tsuchiyama identifies herself not only as a researcher, but as one of the group she is researching, particularly the second-generation Japanese-Americans who were largely "pro-American" and who resented the authority and control of the first-generation Japanese, or *Issei,* at the camps. In this passage, Tsuchiyama is able both to write herself into the "ethnographic present" and to place herself outside of that present in order to reflect on the significance of her experiences. Of course, this rare sortie into personal reflection serves primarily to highlight Tsuchiyama's identification with American values and against any kind of pro-Japanese sentiment.

Overall, moments of obvious personal reflection are rare in Tsuchiyama's reports. Her letters to Thomas provide a much richer account of her own experiences and thoughts—particularly her resistance to the conditions at the Poston camp and her difficulties with the research and writing Thomas required. In one of her first letters to Thomas after moving to Poston from the relocation camp in Santa Anita, California, Tsuchiyama writes: "upon arrival I discovered that many of my notes had been confiscated during the baggage inspection by the police. Before an evacuee leaves for a relocation center his baggage is taken to police headquarters four hours before departure and thoroughly inspected for contraband, and the victim has no knowledge of what has been confiscated until he reaches his destination. I managed to carry out my diary in my purse so from it I have been able to reconstruct as accurately as possible the life at Santa Anita" (Tsuchiyama to Thomas, 24 August 1942). Unlike her Caucasian counterparts, Tsuchiyama was herself an evacuee, and was subjected to the same kind of treatment as other camp residents. Inspections by the FBI or the police were a regular part of camp life, and, according to Tsuchiyama, evacuees could be punished if they were caught with prohibited items such as knives, liquor, electric stoves, scissors, nail files, buckets, tubs, saws, knitting needles, and cash (qtd. in Hirabayashi 39). This letter suggests that Tsuchiyama's own research and writing may have been hindered by her subject position as "Japanese-American."

Ultimately, it seems likely that the rhetorical strategies Tsuchiyama employed—the passive voice, indirect reporting, the chronological organization, and the very sparing use of "I"-oriented discourse—might have contributed to Tsuchiyama's own sense that, in reading her reports, she was viewing life at Poston "from a telescope." She felt temporally, spatially and personally distanced from the events and people she constituted through her writing. Surely, many other factors explain Tsuchiyama's unhappiness with the project, including the physical discomforts of camp life, the lack of resources and administrative support, her fears for her personal safety, her own inexperience with fieldwork, and Thomas's failure to acknowledge these difficulties. Nonetheless, the rhetorical features of "temporal distance" did not provide Tsuchiyama with an opportunity to explore these experiences or even to validate them as part of her research. Hankey faced similar struggles—with the crucial difference being one of ethnic identity—but dealt with those struggles in a different way.

Rosalie Hankey: Immersing Identities

Hankey later described herself as a conscientious and naive student of anthropology when her wartime work began (Wax, "Twelve Years" 133). Like Tsuchiyama, she came from a humble background; Hankey notes in her 1971 book, *Doing Fieldwork,* that she lived in a Mexican-American slum in Los Angeles during the Depression and made a living working on Works Progress Administration (WPA) projects (Wax, *Doing Fieldwork* 64). When Pearl Harbor was bombed, Hankey was in her first semester of graduate work at the University of California at Berkeley. After being rejected from the Waves (the women's branch of the Naval Services) for her poor eyesight, Hankey signed up to work on the JERS study (*Doing Fieldwork* 65). She arrived at the Gila camp in California in July, 1943, and conducted research there and at the Tule Lake internment center in California ("Twelve Years" 133).

Hankey found that many aspects of her identity—race, gender, ethnicity—became salient at different points in her field research. Like Tsuchiyama, Hankey struggled with her first experience as an ethnographer. First, as a Caucasian, Hankey was separated spatially from the evacuees and found it difficult to gain their trust. She wrote later on that "it took me four months to develop any kind of social relationship, not because the Japanese Americans or I were socially maladept, but because almost everyone automatically defined me as 'a spy for the administration'" (*Doing Fieldwork* 18). In a letter of July 12, 1943, Hankey writes to Thomas that she was "warned to stay out of both Caucasian and Japanese camp politics" (Hankey to Thomas, 12 July 1943). Hankey gradually resolved this problem by beginning a series of surveys on mundane topics such as how evacuation had altered the evacuees' way of life, how parents thought evacuation had affected their children, and so on (*Doing Fieldwork* 75–76). These surveys helped her to make acquaintances among the evacuees and eventually form relationships with some of them.

Second, as a young woman, Hankey could cultivate relationships with Japanese-American evacuees only in specific contexts. Like Tsuchiyama, Hankey found that her gender was called into play. She later recalled that the second person she visited at Gila had invited her into his apartment to meet his wife, "But when I arrived, he sent his wife out on an errand and began to pat me on the rump" (*Doing Fieldwork* 69). Hankey writes that she encountered some unsavory Caucasian administrators, including one "old goat" who kept "leering at me and suggesting that a trip to Phoenix might be relaxing" (Hankey to Thomas, 17 July 1943).

Despite these seeming disadvantages, Hankey's race and gender also provided certain advantages. For instance, she was not placed under the same pressure as Tsuchiyama, who needed to guard carefully against any suggestion that she identified with pro-Japanese or anti-American sentiments. Indeed, Hankey freely expressed her sympathy with those who were resisting the United States' violation of the rights and freedoms of Japanese-Americans. Further, as an American of German and Scandinavian descent, Hankey found that her ethnicity sometimes helped allay the suspicion of the evacuees; in one case, an informant deemed her a "German *Nisei*" and therefore someone he could confide in (Wax, *Doing Fieldwork* 53).

While Tsuchiyama felt the need to assume a modest persona in keeping with societal expectations for women, Hankey seems to have had more leeway. She wrote to Thomas that she could use the leering Caucasian staff member to her advantage: "Handled correctly, he might make it possible for you to get the kind of data you want" (Hankey to Thomas, 17 July 1943). Further, to combat the "barbed wire blues," Hankey writes, she developed the following method: "Go to P.X. drink 5 bottles of beer, eat three nut candy bars and a box of pretzels. Discuss the fine points of pool with the head Fireman and G. Brown and the postmaster. Come home with reputation completely ruined, fall into bed. Get up early next morning feeling swell. It works every time" (Hankey to Thomas, 25 July 1943). In her letters to Thomas, at least, Hankey cultivated a kind of "tomboy" persona, perhaps in order to persuade Thomas that she was doing her job well despite the difficulties she faced.

Hankey also played upon feminine stereotypes during her research in the field. In her field notes, dated February 2–3, 1943, Hankey describes how she outwitted a fellow researcher, Opler, because she was concerned that he would try to get her to divulge confidential information she had gathered: "I decided that I would feign disappointment at my progress and play the part of a discouraged, puzzled, female anthropologist, whose hopes to get information had been frustrated [. . .] it worked better than I had anticipated. When I said that I had received almost no information he looked very satisfied and said, 'Now I know by that statement that you're a good field worker.' Thereupon he proceded [*sic*] to show me how vast his knowledge of the situation was" (Hankey *Field Notes*). By playing the innocent, naive student, Hankey got the information she needed.

For all of these reasons, Hankey wrote many years later that her adjustment to life at Gila was long and difficult: "I often felt like a mental defective, and for about six weeks I felt as if I were losing my mind"

(Wax, *Doing Fieldwork* 18). As a novice field researcher, Hankey had to learn the methods of ethnography on the ground. For instance, she wrote to Thomas on July 17, 1943, asking for some examples of "how a case study should be made," since this had been "one of the things for which my anthropological training (up to this point) has not fitted me" (Hankey to Thomas, 17 July 1943, 7). In the same letter, Hankey writes that she was keeping a "detailed journal" while pursuing several major developments: the cooperative at Gila, family life, and developments in relocation, segregation, and the constitution (8). Nonetheless, she seemed unsure about how to shape the results of her observations into a report. In addition to asking for an example of a case study, she writes tentatively: "I shall submit a plan for studying the family in a week or ten days. I want to compose one and try it out before I commit myself" (7). In response, Thomas wrote that she would send along some sociological books, and provided the following advice—which I will quote at length because it provides some indication of the genre expectations cultivated as part of the JERS study:

> Case histories will have to be built up on a long-term basis, sometimes by interview, sometimes by free association. An outline should be followed, or at least used as a guide to check the coverage. Our resettlement outline is a pretty good guide of what we want, and I shall send several copies along [. . .] It is a good idea to use the mimeographed schedule [. . .] as a check on whether you are getting the "factual" or objective background data adequately. The person's "own story" should be told as nearly as possible in his own words. Other persons' accounts of him should be obtained as often as possible. Sources of information should be noted in all instances. Various checks (direct but mostly indirect) should be used to verify the information. There's both art and science in getting a good case history, but mainly you'll have to develop your own techniques and find your own way around. (Thomas to Hankey, 24 July 1943, 1–2)

Thomas's advice seems primarily aimed at ensuring the objectivity of Hankey's accounts, through guidelines, schedules, and cross-checking results.

Perhaps because of her initial difficulties, Hankey reacted by immersing herself into life at the camp, disconnecting herself from the rest of the nation: "During my first difficult months at Gila I threw myself so furiously into my attempts to 'get data' that I almost forgot the existence of the world outside the center and outside the interests of the Evacuation and Resettlement Study. When the American army invaded Italy, I did not hear the news until a Japanese acquaintance told me about it. This, I realized, was going too far, and I subscribed to a daily newspaper and forced myself to read it at my solitary meals" (*Doing Fieldwork* 44). While Tsuchiyama's reports strive for a temporal distance and objectivity, Hankey struggled to gain proximity. As a Caucasian researcher, a novice anthropologist, and a woman, she may have felt particularly pressed to demonstrate her own presence as a researcher in the camp, one who had gained access to the thoughts and opinions of the evacuees.

Perhaps for this reason, Hankey emphasizes her own "ethnographic presence" throughout her writing. "Ethnographic presence" functions as a powerful rhetorical trope, one that helps to guarantee the authenticity and verisimilitude of the ethnographers' account. Clifford Geertz writes that the persuasiveness of anthropological accounts depends mainly on "their capacity to convince us that what they say is a result of their having actually penetrated (or, if you prefer, been penetrated by) another form of life, of having, one way or another, truly 'been there'" (4–5). In Hankey's case, ethnographic presence helped her to construct persuasive field reports, especially since she was not a Japanese-American and had a tough time connecting with her research subjects. Yet, her use of ethnographic presence also got her into trouble with Thomas, who feared that Hankey was losing the objective perspective she thought necessary for good research.

Perhaps the most striking difference between Tsuchiyama's and Hankey's reports lies in the use of the personal pronoun, "I." While Tsuchiyama writes almost completely in the passive voice or by using impersonal constructs, Hankey relies heavily on the first person, and not just as a metadiscursive device. In an early report, *Chronological Account of Segregation,* Hankey reflects on how her own racial identity may have shaped her interviews: "As I was taking down these accounts I noticed on several occasions that the fluency of my informant was increased by the fact that here at last he had a chance to tell a Caucasian just what he thought about the way he and his race had been treated, without the inhibiting effects of the knowledge that his statements would be used as evidence of pre-axis leanings" (*Chronological Account* 42). In this passage, the first person is inextricable from the ideational content of the passage. That is, the "I" does not serve simply the metadiscursive function of facilitating a writer-reader interaction or signaling Hankey's interpretations; these interpretations are central to the analysis of the field events themselves. Later in this report, Hankey describes her own subjective feelings about some of the interviews she conducted: "I had little hope that any would consider talking to me, and, as I began the long hot walk from the Administration Building to Block 39 I had to force down considerable nervousness" (45).

Further, Hankey did not seem compelled to quell her dislike of the WRA or of some of the administrative staff at Gila. From many of her reports, her empathy for the evacuees is readily apparent. In an account entitled *Segregation at Gila* she writes: "No just perspective of this situation can be gained without considering the past experiences of the evacuees. They were uprooted from their homes on the coast, told that they must be evacuated for their own protection. They were again uprooted from the assembly centers. Within recent months extreme pressure has been brought to bear to induce them to leave the comparative safety, security, and comfort of the Relocation Center, and go again to some unknown spot on the outside far from any friends and relatives" (7). In another account, after describing an ascetic meal she shared with evacuees in the mess hall, Hankey remarks: "If I were the mother of children in Gila who were forced to eat a dish like this I would tear K_____ limb from limb" (*Conciliation Begins* 12).

Hankey's writing approaches the style of "vulnerable writing" anthropologist Ruth Behar espouses—a style of writing that not only uses the personal voice, but also focuses on how aspects of one's own self filter perception of the topic being studied and analyzes the various connections between the observer and the observed, the emotional and the intellectual (*The Vulnerable Observer* 13–14). This vulnerability becomes central to Hankey's argument in her reports—an argument for understanding the anti-American expressions of evacuees based on their history of persecution and discrimination, and their poor treatment in the internment camps.

Evidence of a more metadiscursive "authorial presence," as opposed to ethnographic presence, is also evident in Hankey's reports. She sometimes adds reflexive comments on events that she described, as in this selection from her report on segregation: "Viewed from the perspective of the passage of a month's time, the surprise and concern of some members of the administration is amusing" (*Segregation* 5). In other cases, she makes more overt statements to support the credibility of her research: "Realizing the importance this incident would assume I determined to get a correct sequence of the events which had led up to the difficulty and then discover, if possible, why the mess-supervisors had reacted as they did. To obtain a correct sequence of events was extraordinarily difficult because almost every evacuee [. . .] told a story which differed in some respects. By a comparison of reliable evacuees' statements, some of which are included at the back of this report, and the statements of several members of the administrative staff, I trust that I was able to gain a fairly accurate picture" (*Threatened Strike* 2). In other cases, Hankey even admits that her perspectives changed as she gained more knowledge of the evacuees; she writes in one report that "I was a little appalled by the idiotic tone my questions make in transcription and shall be a little more careful hereafter" (*Synthesis* 1). In these examples, Hankey seems to recognize that "temporal distance" was necessary to help her understand the ways in which her own viewpoint may have shaped her research. However, this act of temporal distancing is coupled with subjective interpretations. By writing about her personal reactions to camp events in her early field notes, Hankey could later analyze her notes with an eye to how her own subjective perspective may have influenced her interpretations.

Throughout her wartime writing, Hankey created a sense of ethnographic presence and minimized the kind of temporal distance that seemed to frustrate Tsuchiyama. Hankey's use of the personal voice helped her to reflect on her own position in the field, and by describing some of her emotional reactions and subjective interpretations, she was able to argue for greater understanding of evacuees who were maligned as anti-American. Hankey achieves ethnographic presence rhetorically, through specific discursive choices, but this ability was also enabled by her position as a Caucasian woman, one who might have felt less susceptible than Tsuchiyama to questions about her own loyalty to America.

Thomas strongly reprimanded Hankey for her failure to assume an appropriately distanced and objective voice in her writing. In a letter of November 10, 1943, Thomas sent along her comments on Hankey's first report on Segregation. In addition to critiquing its organization and coherence, Thomas writes that "[t]here is evidence of a definite bias on your part in two respects: (1) your contempt of most of the Caucasians, and (2) your pro-Issei and anti-Nisei bias which comes out very clearly at several points" (Thomas to Hankey, 10 November 1943). Thomas critiques Hankey's tendency to call people "dumb, stupid, silly, and so on, which is all right in letters but just doesn't work in reports" (ibid.). Thomas advises Hankey to "Isolate your own interpretations as much as possible" (3), "don't talk about yourself" (3), and "[a]void name calling and let the Caucasians' and evacuees' behavior speak for itself" (4). In response, Hankey agreed to revise the report based on Thomas's comments and to stifle her own subjective reactions: "My 'Caucasian contempt' I fear I cannot lose, though it is not proper that it appear in the reports. I shall carefully delete it and refrain from expressing it in later reports" (Hankey to Thomas, 15 November 1943).

A later letter suggests that this "problem" only worsened before it improved. As Hankey became more and more immersed in camp life,

Thomas continued to be concerned with the tone of Hankey's reports and letters. In one undated (but presumably later) letter, she writes:

> Your letters and recent reports give me a strong impression that you are falling into a state of mind that may, unless it is checked, seriously bias your reports. You are, it seems to me, over-identifying-yourself [*sic*] as a part of the "core of the Japanese community." [. . .] Now, Rosalie, you can never become a Japanese or instead of "never," let us say at least not in time to do the study any good. As an anthropologist, you know damned well assimilation or acculturation or whatever you want to call it is not achieved by a process of swift conversion. I don't want you to carry this identification to the point where you lose the objectivity of the scientist. (Thomas to Hankey, undated)

Thomas warns Hankey that, unlike Tsuchiyama, she will always be "on the outside looking in," no matter how close a rapport she could strike up with the evacuees.

Under Thomas's tutelage, Hankey seems to have learned to assume the appropriately distanced objectivity required of a scientist and to enact that objectivity through her writing. Hankey wrote that she would make more of an effort "to appreciate the Nisei and the Caucasians objectively [. . .] I shall appreciate detailed criticism of any biased remark I make in the future. I assure you that I am not trying to become a Japanese [*sic*]" (Hankey to Thomas, 17 December 1943). On the same day, she sent Thomas her next report, under separate cover, with the message: "Here it is. I tried to be careful to keep free of prejudice. [. . .] My own and the Kondo's reactions to the hearing may be superfluous; if so, please advise me and I shall omit such things in future" (Hankey to Thomas, 17 December 1943). This report, according to Thomas, was "excellent," and apparently free of the subjective bias found in the earlier reports (Thomas to Hankey, 30 December, 1943). Hankey wrote later that she felt satisfied with the report: "For the first time I felt that I had gained some approximation of a balanced account" (Hankey to Thomas, 1 January 1944). Apparently, Hankey continued to take this critique to heart. She wrote of her report on Tule Lake that "I went over the first copy, after having studied the criticised [*sic*] segregation report thoroughly, and slashed out all person references I could find" (Hankey to Thomas, 26 February 1944). This report, too, was also apparently satisfactory; Hankey wrote to Thomas that "I am very cheered by the fact that my report was not too bad. I know that future stay in Tule will enlarge, correct and bring it into better perspective" (Hankey to Thomas, 5 March 1944). Hankey's example shows how learning a genre also involves learning to assume a certain kind of subjectivity.

Dorothy Thomas: Erasing Identities

The previous analysis suggests that both Tsuchiyama and Hankey, as young anthropologists, learned to assume temporal distance and objectivity as rhetorical strategies and as attitudes toward their research. By coaxing her field reporters to write in this manner, Thomas ensured that she could easily integrate the reports she received into the larger synthetic account she was producing. This section considers how the books Thomas wrote based on the JERS study erased Tsuchiyama's and Hankey's subjective responses to camp life, and capitalized on the objective field reports these women, and other researchers, produced.

In addition to the temporal distance she required in field reports, Thomas enacted other rhetorical practices to ensure the smooth upward flow of information. She made it a practice to have her field researchers critique each other's reports, a decision that seems to have generated some competition among her young researchers (Hirabayashi 80). She asked Tsuchiyama to visit Gila, where Hankey was working, but warned Hankey in advance that Tsuchiyama was "one of the most talented members of our staff but, between us, the most neurotic" and a "terrific prima donna" (Thomas to Hankey, 1 September 1943).[5] Meanwhile, Tsuchiyama and Nishimoto wrote a report on Gila based on their visit, a report that was partially meant to check up on Hankey's research. Hankey was given the opportunity to read this report and make her own corrections and comments. Hankey wrote to Thomas: "I think that Tamie and X. are slightly obsessed with the 'evil intentions' of the Evacuees, and are inclined to over-emphasize their mercenary and subversive qualities" (Hankey to Thomas, 20 October 1943). These comments were then passed on to Nishimoto—something that led Hankey to worry that Nishimoto would "be through" with her (Hankey to Thomas, 1 November 1943). While "peer-reviewing" reports in this manner probably helped to ensure their accuracy and quality, it seemed to have generated tensions between the researchers. As the only two female field researchers on the study, Tsuchiyama and Hankey could have formed a support network for each other, but were instead placed in a competitive relationship—in part due to Thomas's caustic comments about Tsuchiyama.

Thomas's own background in statistical sociology also shaped her reactions to reports written by the anthropologically trained Tsuchiyama and Hankey. In one letter to Hankey, for instance, Thomas warns against "one of the anthropological fallacies, that is, in assuming a uniformity of reaction and expression" based on only a small sample of informants. Thomas claims that anthropologists were "accustomed to deal with the static rather than a dynamic situation," and that they "depend far too much on a limited number of informants, often only one, who gives a picture of a situation which is bound to be individual, but which the anthropologist too often takes as representing universals" (Thomas to Hankey, 15 November 1943). Rather than the in-depth view of the field that anthropologists provided, Thomas seems to have preferred a "big picture" account of internment as a whole. For this reason, the detailed records kept by the informants were not valued in and of themselves, but were required only so that Thomas could construct a broader account of the phenomenon.

Ultimately, Thomas drew from Hankey and Tsuchiyama's notes and reports, along with those of her other field researchers, for *The Salvage* and *The Spoilage.* Yet, both books are very much positioned as Thomas's work, first and foremost. *The Spoilage,* which analyzes those evacuees who "failed" to assimilate into American culture, lists Richard Nishimoto as coauthor, and credits Rosalie A. Hankey, James M. Sakoda, Morton Grodzins, and Frank Miyamoto as contributors. Tsuchiyama's name is not listed on the frontispiece among these contributors, but two of her reports are excerpted in the text and acknowledged in footnotes. Further, anthropologist Lane Ryo Hirabayashi argues that *The Spoilage* "is actually largely based on Rosalie Hankey's field notes" (81), but the extent of this contribution is not clear from the acknowledgments in the book itself.[6]

Throughout the book, Thomas and Nishimoto often quote statements taken by a field researcher, such as transcriptions of speeches or verbatim statements by evacuees. These passages are credited with the name of the evacuee or a description (i.e., "Statement by Mrs. Tsuchikawa," "Statement by a young Nisei girl from Topaz"), along with the designation "field notes" and the date (305), but do not provide the name of the researcher who recorded the statement. Thus, Thomas and Nishimoto extract the "original data" out of the field reports and largely erase the analytical voice of the field researcher. For this reason, it is not immediately apparent to whom the field notes belonged, although one can assume that they were not Thomas's, since Thomas did not herself conduct any of the field research.

The Spoilage covers the experiences of those Japanese-Americans of the first generation who returned to Japan, and of those of the second generation who renounced their American citizenship (xii). Overall, *The Spoilage* functions mainly to support a preexisting theory: an assimilation model of Japanese-American identity, based on social demographics, and an assumption that those who returned to Japan ultimately "failed" to assimilate. The book includes thirteen chapters, organized along a roughly chronological scale, beginning with Evacuation (Chapter 1), Detention (Chapter 2), Registration (Chapter 3), and Segregation (Chapter 4), and ending with Resegregation (Chapter 12) and Renunciation (Chapter 13). Thus, it charts the "progress" of evacuees from the time they were first removed from their homes, to the final stages of resettlement or renunciation.

Although published in 1946, not long after the war, the book situates the events of relocation and resettlement fully in the past. For the most part, it reads as a historical narrative of the major events as they occurred across the different internment camps. For instance, Chapter 10, titled "Interlude: Period of Apathy," begins as follows: "The period from the end of April to the middle of June, 1944, was, politically and psychologically, characterized by indifference and apathy on the part of the residents. The protest movement had failed, and its leaders were still confined in the stockade. The Army had taken over and then relinquished control of the center. A partial strike had been succeeded, if not by 'full employment,' at least by a wider distribution of jobs, and wages were being paid" (236). Thomas's use of the omniscient, third-person style increases the apparent objectivity of the account, while the specific contextual details provided by field researchers are subordinated to larger cultural patterns and historical events. By erasing the voices of the individual researchers, Thomas and Nishimoto craft an "objective" account of the entire process, but in doing so, they fail to provide the context for the individual events or remarks that they record.

The next book in the series, *The Salvage* (1951), lists Charles Kikuchi and James Sakoda as coauthors. As with *The Spoilage,* this book is organized around a predetermined theory, one that assumes the desirability of "social integration of *Issei* into the larger American community" (133). The first part of the book includes chapters on various cultural patterns or topics that help to explain the successful assimilation of Japanese-Americans: "Demographic Transitions," "Urban Enterprise," "Occupational Mobility," "Religious Differentials," and so on. The second part includes fifteen "life histories" of individuals, including a "Schoolboy," "Agricultural student," "Journalist," "Counter Girl," and others. These case histories were conducted by Charles Kikuchi between April, 1943 and August, 1945, based on an extensive outline covering the individual's life history and experiences in the relocation camps (*The Salvage* 136). The evidence gathered from interviews with evacuees helps to reinforce a predetermined set of patterns and considerations, which were embedded into the meta-genre used to coordinate interviews and into the organization of the book as a whole.

The Discourse of Objectivity

Overall, it is possible to sketch out the relationships between institutional arrangements of space and time, genres, and the discourse of objectivity. As written texts moved further away from the field (from field notes and letters, to reports, to the published books), they became more "objective" and more distanced, temporally and spatially, from the events they described. The books Thomas published as director of the project garnered the most professional acclaim, but did so by disavowing a personal voice. Thomas may have adopted this stance knowing that, as Behar suggests, the work of women who flout the conventions of standardized genres has tended to be denigrated as "confessional" or "popular," rather than "scholarly" or "theoretical" ("Introduction" 4).

Tsuchiyama and Hankey struggled to downplay the subjectivities that emerged during the course of their research in order to produce "objective" field reports. As novice field researchers and anthropologists, both women found it difficult to adjust to the working conditions in the camps. This lack of both practical and emotional preparation made the initial adjustment to camp life difficult for both field researchers. Hankey managed by eating copiously in her first few weeks at Gila and gaining several pounds (her letters to Thomas sometimes include references to pounds gained and lost), while Tsuchiyama struggled with heat rash and exhaustion (her letters to Thomas frequently mention her difficulties adjusting to the desert climate). In their letters to Thomas, both voiced concerns that Behar lists as common during the anthropologists' trip into the field: "Loss, mourning, the longing for memory, the desire to enter into the world around you and having no idea how to do it, the fear of observing too coldly or too distractedly or too raggedly, the rage of cowardice, the insight that is always arriving late, as defiant hindsight, a sense of the utter uselessness of writing anything and yet the burning desire to write something" (*Vulnerable Observer* 3). It seems that neither woman was prepared to deal with these feelings of inadequacy, fear, and loss.

Despite their common standing as novices, however, Tsuchiyama and Hankey were positioned differently by their raced or gendered identities. Hankey wrote in *Doing Fieldwork* that her status as a Caucasian woman sometimes worked to her advantage, and that she cultivated different identities in order to garner information:

> Some Japanese Americans felt more comfortable if they could treat me like a sympathetic paper reporter [. . .] In Tule Lake the superpatriots and agitators found it easier to talk to me once they had convinced them-

> selves that I was a German "Nisei," "full of the courageous German spirit." I found this fantasy personally embarrassing, but I did not make a point of denying my German ancestry. Finally, I was not a geisha, even though a shrewd Issei once suggested that it was because I functioned as one that I was able to find out so much of what happened at Tule Lake. His explanation was that Japanese men—and especially Japanese politicians—do not discuss their plans or achievement with other men or with their wives, but they are culturally conditioned to speak of such matters with intelligent and witty women. (53).

This passage suggests that Hankey cunningly accentuated certain aspects of her identity and drew upon culturally available tropes (the German *Nisei* or the geisha) in her fieldwork. Further, as a Caucasian, Hankey felt free to criticize and even ridicule the WRA administration, empathize with the evacuees, and mention the general discrimination Japanese-Americans faced throughout the country, prior to the war. Thus, her greater freedom in expressing her own opinions in her reports may have to do with her more privileged position as a Caucasian American, one who would did not have as much at stake if she allied herself too closely with pro-Japanese groups in the camp.

Of course, Hankey was not immune to censure for her views. According to Orin Starn, Hankey was one of the only anthropologists working on internment who was openly critical of the WRA and its policies, and she was eventually dismissed for this reason (708). In her own accounts, Hankey writes that she was removed from the study suddenly after Thomas had been warned by a WRA official. Her list of supposed infractions was long: "I had consorted with pro-Japanese agitators and attended ceremonies devoted to the worship of the Japanese emperor. I had had immoral (sex) relations with a number of Japanese Americans. I had made disrespectful remarks about the project director. I had been a general troublemaker and had tried to subvert WRA policies. I was by temperament an anarchist, and, since my mother had been abused by members of the Los Angeles police force, I had no respect for the law. I had communicated with the Department of Justice" (*Doing Fieldwork* 169). According to Hankey, Thomas decided to remove her from the project not because she believed these accusations, but because she feared a scandal would jeopardize the reputation of the study and the eventual reception of the publications ensuing from it. Despite her dishonorable removal, after the war, Hankey went to the University of Chicago, where she completed her Ph.D. in 1950. She went on to have a successful career in anthropology, working alongside her husband, Murray Wax, on studies of American Indian culture.

As a Japanese-American who was required to demonstrate her own patriotism, Tsuchiyama could not appropriately express strong opinions or perspectives in the same way that Hankey could. Indeed, among her field reports, the moments in which Tsuchiyama's subjectivity comes to the fore are mainly 1) a passage that relates patriotic feelings at the sight of the American flag, and 2) passages where Tsuchiyama expresses her displeasure at the segregationists, those evacuees who remained loyal to the Japanese nation and refused to declare allegiance to the United States. By demonstrating too much identification with the evacuees, Tsuchiyama would be jeopardizing her own tenuous position as an "objective" researcher. Instead, in her writing she enacts temporal distance, a strategy that not only fulfilled Thomas's expectations, but perhaps also functioned as a coping mechanism. Indeed, Kirsch has found this strategy to be quite common among the academic women writers she has studied (64–65).

Ultimately, Tsuchiyama also left the JERS study, after a series of increasingly tense exchanges with Thomas. Initially, Tsuchiyama left Poston in October of 1943 for Cleveland (and then Chicago), where she hoped to complete a series of reports based on her extensive field notes. On January 24, 1944, Tsuchiyama wrote to Thomas, saying that she would finish up her reports and then afterward, "thrown [*sic*] in the sponge," claiming "I don't want to deal with human beings again in my life. I have reached the conclusion the Japs in Poston do not care to be helped, do not want to be" (Tsuchiyama to Thomas, 24 January 1944). Tsuchiyama was despondent for a number of reasons, one being that she had been accused by Nishimoto's wife of having an affair. This was despite Tsuchiyama's belief that "it was quite evident to everyone that I had no more intentions on capturing X [Nishimoto] than marrying a hippopotamus" (Tsuchiyama to Thomas, 24 January 1944). Thomas assured Tsuchiyama that her disgust with people was understandable, given the circumstances, that she had "the greatest admiration for your abilities and contribution," and that she should continue on in Chicago and would be free to leave the study once her report was finished (Thomas to Tsuchiyama, 26 January 1944).

The report never materialized, despite assurances by Tsuchiyama that she had every intent to finish it (Tsuchiyama to Thomas, 31 January 1944 and 12 July 1944). By July, Tsuchiyama wrote to Thomas saying that her progress had been slow, a fact she attributed to "my having become 'stale' from being in contact with the same subject matter too long" (Tsuchiyama to Thomas, 12 July 1944). Thomas replied that she was "disturbed" by Tsuchiyama's lack of progress, and asked for an incomplete version of the report and whatever notes she had ready (Thomas to Tsuchiyama, 15 July 1944). Tsuchiyama agreed to this request, but not to Thomas's suggestion that she come to Berkeley for a time to go over the material. Tsuchiyama's dissatisfaction is obvious:

> Your letter of July 15 clearly indicates that you do not trust me and feel that a constant surveillance is necessary. Since it is impossible to me to work under the conditions you set forth, and since it has been my policy as long as I can remember never to work for an employer that distrusts me, I am resigning from the Study as of July 15, 1944. [. . .] I consider a trip to Berkeley wholly unnecessary. From an ethical standpoint, I feel I am under no obligation to do so, and from the practical point of view, I believe that my field notes are sufficiently full—perhaps too full—for an armchair social scientist to evolve fanciful theories. I am sorry that I have so little completed on paper [. . .] but I am certain you or X can do a much better job in a much shorter time. (Tsuchiyama to Thomas, 17 July 1944)

After Thomas wrote that Tsuchiyama's reports lacked organization, context, and detail, Tsuchiyama requested once again to leave the study, a request that Thomas granted as of August 2, 1944 (Thomas to Tsuchiyama, 2 August 1944).

Tsuchiyama went on to join the Women's Army Corps, and was assigned to Military Intelligence Service, where she did translations from Japanese to English (Hirabayashi 153). She finished her Ph.D. in anthropology in 1947, but (perhaps due to racial discrimination), could not find a suitable academic position. Tsuchiyama never published her research on Japanese-Americans and ultimately left the profession. She conducted research in Japan under the U.S. occupation forces, and then finished her career as a librarian (Hirabayashi 157–59).

Speaking of genre, Charles Bazerman writes that "insofar as each text and each genre implies a set of relations, recognized social positions and roles, stances of cooperative or competitive work, and sets of relations, recognized social positions and roles, stances of cooperative or competitive work, and sets of typified discursive relations, learning to formulate statements of the accepted genres integrates the novice into roles and positions within structured relations" ("Discursively" 304). As novices, Tsuchiyama and Hankey gradually came to acquire the generic tools and knowledge required to conduct the sort of ethnographic research Thomas preferred. In the process, they were also interpellated into the structured roles and positions required of novice field researchers within a hierarchically organized set of genres. Yet, their success in this regard carried a price. The genres privileged in the JERS discouraged Tsuchiyama and Hankey from writing field reports about their subjective experiences at the internment camps, or how their subjectivities were bound up with those of the internees, or how that might have affected their perception of camp events. Indeed, some critics have argued that anthropologists studying internment during World War II contributed to a discourse that "preempted criticism of removal and legitimized the internment" (Starn 702). With its emphasis on distanced, objective accounts, the genre of the field report encouraged this lack of criticism of internment.

More broadly, the genres used in the JERS militated against what feminist researchers have called "situated knowledges" (Haraway, *Modest Witness* 11). Hankey and Tsuchiyama's letters and reports suggest that both anthropologists had considered how their own subjectivities shaped and were shaped by their research on the JERS study. However, by the time their reports made their way into Thomas's books, those reflections were omitted. In this way, Thomas excluded the critical perspectives that have been advocated by Linda Brodky, Eleanor Kutz, and other scholars who argue for, in Kutz' words, ethnographies that not only tell us *what* the ethnographer sees, but "*how* she is seeing [. . .] calling our attention to the ways in which her point of view shapes what she sees, and thus what we see" (343–44). By erasing this critical perspective, the "discourse of objectivity" ultimately prevented the more critical insights Hankey and Tsuchiyama might have offered.

3. *Women Physicists on the Manhattan Project*

> I think everyone was terrified that we were wrong (in our way of developing the bomb) and the Germans were ahead of us. That was a persistent and ever-present fear, fed, of course, by the fact that our leaders knew those people in Germany. They went to school with them. Our leaders were terrified, and that terror fed to us. If the Germans had got it before we did, I don't know what would have happened to the world. Something different. Germany led in the field of physics, in every respect, at the time war set in, when Hitler lowered the boom. It was a very frightening time.
>
> —Leona Woods Marshall

In 1942, the escalating war prompted a massive undertaking on the part of scientists, engineers, technicians, and military officials to develop the first atomic bomb, a device that would offer unprecedented destructive power. Experts predicted that a ten-kilogram atomic bomb would release the same amount of energy as 1,000 or 2,000 tons of TNT (Hoddeson et al. 22, 28). Fearing that the Germans were developing a similar weapon, physicists in the United States pressured the government to hasten research on the bomb. Whoever finished first would have a tremendous advantage—enough to win the war.

The Manhattan Engineer District, organized officially in 1942, eventually employed 130,000 people in an all-out race to develop the device before the Germans (Hughes 9). Recent studies of scientific laboratories suggest that a similar urgency may drive everyday research today, especially when deadlines for conferences or publications are looming. Latour and Woolgar write that "The tension of a battalion headquarters at war, or of an executive room in a period of crisis does not compare with the atmosphere of a laboratory on a normal day!" (229), while Elinor Ochs and Sally Jacoby argue that tight timelines help to drive physicists toward agreement on what data means and how it is to be interpreted (Ochs and Jacoby). Clearly, the urgency attached to scientific research can lead to productive consensus and new discoveries. Some even credit the Manhattan Project for inaugurating this intensive approach to scientific research.

However, as I will show, the Manhattan Project ushered in a distinctively single-minded approach to scientific research, one that maximized productivity and competition at the expense of critical reflection and productive conflict. Examining this approach is important because the Manhattan Project has come to represent the "gold standard" for high-paced scientific research, and it now serves as a metaphor for any intensive scientific or technical project. For example, in 1993, President Bill Clinton called for a "Manhattan Project" on AIDS research (qtd. in "A Renewed Urgency on Aids" A32). In 2005 the Department of Energy called for a Manhattan Project on solar energy (Service 1391), while that same year the United States government funded a twelve million dollar research project on terrorism deemed "the social science equivalent of the Manhattan Project" (Holden 511). Although this intensive approach to scientific research may be tremendously productive, enabling technological developments to proceed at breakneck speed, it also constrains the rhetorical options available to scientists and may prevent them from voicing ethical or social concerns.

The intense pressure to produce the atomic bomb led scientists to limit their focus to technical concerns and to frame issues of ethics, health, and safety in narrow, quantitative terms. One Manhattan Project scientist, J.H. Manley, writes that "Under the pressure of a wartime objective, there was little inclination for more than a passing thought about what man would ultimately do with this new elemental source of energy for constructive or destructive purposes" (139). Another scientist, Frederic de Hoffman, wrote after the war that ethical questions did not occur to him or his fellow scientists: "We were merely concerned with the project and the driving urge to get to our wartime goal as fast as possible" (168). Those scientists who worked with safety hazards did so within a limited technical structure, because broader ethical concerns were excluded from decision-making frameworks and genres such as technical reports. This technical rationality—a focus on narrowly defined, quantifiable criteria—constrained scientists' efforts to consider the broader implications of their research on the bomb.[1]

While it is not necessarily true that women, as a group, always demonstrate a greater concern for ethical issues in science, it has been suggested that a *feminist* form of science would place greater emphasis on ethics and sustainability (see Schiebinger, "Creating Sustainable Science"). In this chapter, I examine how two women scientists handled ethical concerns about safety during their work on the Manhattan Project. Leona Marshall and Katharine Way have received parenthetical mentions in existing scholarship on the Manhattan Project, but the details of this work remain unclear. Using declassified reports analyzed here for the first time, I examine how these women adapted to the dominant discourses of urgency and technical rationality.

Marshall developed time-saving methods for calculations and prioritized speed of production, framing her calculations so that they supported the overall narrative of technical rationality and efficiency. Marshall internalized these values to such an extent that she not only incorporated them into her technical reports, but also her lived experience. During the course of her work at Hanford, Marshall became pregnant with her first child. She worked in the plant until two days before she gave birth, and returned to work just one week later, not wanting to miss out on the scientific work being done there. After the war, Marshall became an advocate and lobbyist for peaceful uses of nuclear energy.

Way was situated within the same context of urgency as Marshall was; however, her reports and memos indicate that she was concerned about the safety implications of nuclear research. Nonetheless, her rhetorical choices were constrained by genres that admitted only technical considerations, and she was ultimately unsuccessful when she tried to raise questions about human safety in the plutonium plants. After the war, Way became involved in a scientists' movement for cooperative, international control of nuclear technology.

Although Marshall and Way each reacted differently to the rhetorical constraints they faced, both examples suggest that the intensity and urgency of the Manhattan Project limited scientists' ability to address the ethical implications of their work or to engage in productive conflicts over the values shaping that work. In this chapter, I examine how the discourses of urgency and technical rationality shaped the organizational culture of the Manhattan Project, how Marshall and Way were positioned within that network, and how these factors impinged on their rhetorical options in internal memos and reports. Ultimately, this example indicates that when technical rationality is endemic within a scientific organization, it may limit discourse about ethical issues, including human and environmental health.

The Gendered Culture of the Manhattan Project

Brian Easlea has argued that the nuclear arms race has been essentially a masculinist endeavor, one undertaken primarily by men in "a paradigmatically masculine spirit" (7). Easlea cites both the sexual and birth metaphors often associated with the development of nuclear technologies, and the intensely competitive nature of nuclear research (41). While many women did participate in nuclear research during World War II in America, that research was constrained by a culture that encouraged urgent competition to complete the atomic bomb, not careful reflection about the effects of nuclear technology.

The Manhattan Project included three main phases of work conducted at four core locations, although a number of university research centers and industrial groups were contracted to assist in the Project.[2] First, at the University of Chicago's Metallurgical Laboratory (called Met Lab), founded in January, 1942, scientists worked under Enrico Fermi to develop the first self-sustaining nuclear chain reaction. Research could proceed on the atomic bomb only if scientists could prove that a nuclear reaction was physically, not just theoretically, possible.

Second and third, at Hanford Engineer Works near Richland, Washington and at Clinton Laboratories in Oak Ridge, Tennessee, scientists and engineers set up full-scale nuclear reactors and developed methods to produce the quantities of enriched plutonium and uranium necessary to build the actual bomb. The Hanford plant was managed by the DuPont company, whose engineers and managers imposed a strict time schedule. In fact, the time pressure at both of these locations was high: Scientists and engineers had two years to begin producing quantities of uranium and plutonium, but after that, officials believed, "every month's delay would have to be counted as a loss to the war" (Hawkins 17). By December, 1943, only two milligrams of plutonium had been produced, so it was important to keep production on pace throughout 1944 and 1945 (Groves 41).

Fourth, at Los Alamos, New Mexico, a secret community of scientists, technicians, engineers, and military personnel designed and built the atomic bombs with plutonium and uranium produced at Hanford and Oak Ridge. The workers at Los Alamos included female scientists, but also Women's Army Corps (WAC) members and the wives of scientists who were pressed into jobs to help speed the Project along.

Despite the urgency of the Project and the need to maximize output—or perhaps because of it—women's positions generally followed a pattern of hierarchical segregation. Although a select group of male theoretical physicists is generally credited with the achievement of the atomic bomb—men such as Robert J. Oppenheimer, Fermi, and Leo Szilard—the Manhattan Project as a whole was a much larger undertaking, and women contributed to the scientific work at each of the four locations. Women comprised 9 percent of the 51,000 employees at Hanford in 1944, and 30 percent of the employees in the Tech Area at Los Alamos (Howes and Herzenberg, *Their Day in the Sun* 13–14). However, most women, even women with a Ph.D., were assigned to lower-level tasks, such as performing calculations.[3] For example, physicist Jane Hamilton Hall had just finished a Ph.D. at the University of Chicago, but she was assigned to write a review of reports concerning safety hazards and calculated the probable dispersion of fission products in the atmosphere—work she considered beneath her level of scientific training. Although most of these women were white, some African-American women (such as Mildred M. Summers) and Native American women (such as Agnes Stroud) also participated (Howes and Herzenberg 123–24, 130–31). The Chinese-American physicist Chien-Shiung Wu conducted research for the Division of War Research at Columbia University (Howes and Herzenberg, "Chien-Shiung Wu" 614). To date, however, the overwhelming majority of the 300 or so women who have been identified as Manhattan Project workers were white, probably because even the exigence of war failed to dislodge the deeply sedimented biases against women of color in scientific disciplines.

Professional status and gender played a role in the kinds of documents women wrote and the rhetorical strategies available to them. In general, the kinds of reports women scientists wrote would have been addressed to superiors, either scientists who were their group leaders, or in some cases DuPont personnel, who would use that data for larger reports or for decision-making purposes. Within this hierarchical organization, women tended to be concentrated at the lower levels.

Marshall was just finishing her dissertation at the University of Chicago when the Met Lab was established there in 1942. She was the only woman present when the initial chain reaction was achieved by Fermi's team in December of that year. Marshall's main task was to construct detectors that would measure the flow of neutrons in experimental atomic piles at the University of Chicago's Metallurgical Laboratory (Howes and

Herzenberg, *Their Day in the Sun* 38). Later, she worked on radiation safety at the Hanford Plutonium Plant.

Way worked with a different group at Met Lab under Eugene Wigner. After graduating with a Ph.D. in physics from the University of North Carolina in 1938, Way taught physics at Bryn Mawr and the University of Tennessee. At the start of the war, she was recruited for a position in Washington, D.C. that involved minesweeping, but it did not hold her interest. Upon hearing rumors of nuclear research going on in Chicago, Way called her graduate advisor, John Wheeler, and asked if he knew of any openings. Within a few days, Way was on her way to Chicago, where she worked on reactor design and analysis of fission products (Martin et al. 573–74). She helped to design the nuclear reactors for Hanford, studied radiation hazards in the plant, and organized radioactivity data on fission products (573–74).

In addition to their various duties, Marshall and Way each researched issues surrounding radiation safety in the Hanford plant. Given the urgency of the Project, their writing was situated within a culture that valued efficiency, speed, and technical rationality. From the start, appeals stemming from government, military, and scientific officials lent the Manhattan Project an air of utmost urgency. The government provided the main impetus for work on the atomic bomb. On March 11, 1942, President Roosevelt wrote to Vannevar Bush, the director of the Office of Scientific Research and Development (OSRD): "I think the whole thing should be pushed not only in regard to development, but also with due regard to time. This is very much of the essence" (qtd. in Stoff, Fanton, and Williams 26). The government stipulated that work on the Manhattan Project was to be conducted under maximum speed and absolute secrecy. In a 1944 interview published in the *New York Times,* Bush stated: "Under ordinary peacetime procedure it takes about five years to develop a new idea and put it in the hands of the people. It would be dangerous for an industrial concern to attempt to do it in less time. In war, things must be done quickly. Yet there is as much danger if we act too hastily as if we delay too long. Practically all our work must be carried on under the greatest secrecy, and that in itself is a handicap to the speed which is essential" (qtd. in Woolf 16). Delay became a danger in wartime—a danger given the same, or possibly more, weight as more common dangers for scientific projects, such as human safety. Given the urgency imposed by the government, the military director of the Project, General Leslie Groves, decided early on that the design and construction process for the production plants would proceed as quickly as possible, even though their work would be based on very limited laboratory data.[4]

The scientists themselves were motivated by the fear that the Germans were working on an atomic bomb, and this fear was especially pressing for those who had emigrated from Europe. Several scientists had themselves urged the United States government to begin research on nuclear weapons. In 1939, a group of prominent physicists, including Albert Einstein, Leo Szilard, and Enrico Fermi, warned President Roosevelt that recent research into uranium raised the possibility for a new type of explosive bomb based on the (hitherto unproven) concept of a nuclear reaction.[5] In a letter of August 2, 1939, the physicists informed Roosevelt that Germany had taken over a series of uranium mines in Czechoslovakia, and insinuated that the Germans had already begun a research program to produce a nuclear bomb. The letter called on Roosevelt "to speed up experimental work" on uranium (Stoff, Fanton, and Williams 19).[6] Several of the leading scientists had recently emigrated from Europe and knew firsthand of the threat the Germans represented.[7] They articulated this threat to others working on the Project. For example, Alvin Weinberg, one of the Met Lab physicists, noted that Szilard and Wigner lived "in mortal terror" of the Nazis, and that their fear spread to others involved in the Project (qtd. in Sanger 40). Scientists reasoned that the Germans would stop at nothing to win the war, and the atomic bomb would make the Nazis nearly unbeatable.

When it became evident later on that the Germans would not succeed in developing an atomic bomb, the argument shifted. Officials began to argue instead that developing the bomb would help to speed the end of the war and save lives. D. M. Giangreco writes that by 1944, one prediction was that invading Japan would cost one million casualties. This figure was cited, for example, by Secretary of War Henry Stimson in a memo to President Truman of July 2, 1945—just two weeks before the successful test of a nuclear bomb at the Trinity Site in New Mexico (561). Officials also expected the Japanese war to last well into the latter months of 1946 (566). Although other figures circulated, this estimate was widely adopted for military purposes. Official estimates were restricted to top-secret planning documents, but Giangreco suggests that the casualty estimates were widely discussed by military officials (535). This forecast may also have circulated among the scientists at work on the Manhattan Project.

Given the sense of urgency that permeated the Project, scientists developed research methods that would increase the speed of their work and, concomitantly, a rhetorical culture that would facilitate that research while simultaneously maintaining secrecy. Scientists who were used to discussing their theories and findings openly struggled to accept a hierarchical, compartmentalized system established by military officials. While the scientists favored communicating informally and horizontally with their team members, the military and industrial officials instituted more formal memos and reports. These documents would circulate only to certain individuals, based on their position and level of security clearance.

The genre of technical report that emerged at the Manhattan Project loosely mimicked that of the scientific report, organized around the IMRD structure: Introduction, Methods, Results, and Discussion (or Conclusions). Writers often placed the conclusions toward the start of the document, perhaps because work on the Manhattan Project was so urgent; this enabled decision makers to find the most important information quickly.

More importantly, as is the case in scientific reports, the criteria used for evaluating courses of action were technical ones, supported by calculations. In other words, social, ethical, and moral criteria rarely figured directly—a problem Carolyn Rude has identified in technical reports more generally (78). Instead, scientists focused mainly on quantitative values such as speed, productivity, and efficiency. As a rhetorical strategy, technical rationality reframes social and moral issues in technical terms, thereby disengaging those issues from the social and political contexts in which they are embedded (Fischer 359). This technical rationality ultimately enabled scientists to exclude the ethical implications of their research on a bomb of unprecedented destructiveness, even when they were dealing directly with safety issues.

However, this does not mean that scientists like Marshall had no rhetorical agency in shaping their reports. Even calculations work, including the choice of starting assumptions, involves rhetorical and ethical decisions. By choosing assumptions that accorded with the dominant technical rationality on the Project, Marshall helped to ensure that her reports were persuasive. Meanwhile, when Way based her calculations on issues of safety, they were dismissed or questioned as "unreasonable." Thus, the calculations scientists made in these reports were not neutral, but depended upon rhetorical judgments as well as scientific ones.

Leona Marshall: Managing Risks

Throughout the Manhattan Project, scientists worked with limited knowledge and experience within a context shaped by two major risks. On the one hand, not pushing plutonium production quickly enough would prevent scientists from finishing the bomb in time to stop the

Germans (or to end the war). On the other hand, pushing production too much would loosen safety standards and threaten the health of workers. Although General Groves wrote after the war that safety policies were strictly followed, he also notes that these procedures "enormously increased the time—which was most important—and the cost" of the Project (Libby 86). Declassified documents show that despite Groves's assurances, scientists sometimes had to weigh timeliness against safety and to take risks based on the relative importance of each of these concerns. In those instances, the most persuasive data tended to be that which enabled the utmost speed of production without sacrificing safety standards. This data was justified through technical rationales rather than ethical ones. Or, more precisely, speed and efficiency—justified by scientific data—became the dominant ethic on the Manhattan Project. Marshall's reports demonstrate how technical rationality functioned as an imperative guiding the design and organization of technical reports, particularly in reports that assessed safety risks.

When assessing risks, one may select certain starting assumptions to make the risk seem more or less acceptable to the audience. For example, if the audience were highly interested in safety, one might calculate a more conservative estimate of risk, assuming that the worst case scenario was indeed a serious possibility. The more conservative the estimate, the lower the level of risk one is willing to accept. If the audience were interested in productivity, one might determine a less conservative estimate. Here, one might accept a somewhat higher level of risk in order to keep production at a higher pace. Of course, presenting risk assessments to an audience also involves rhetoric, since one may either downplay or highlight the risks of a given activity through the language used to describe it. The very term *conservative* takes on a pejorative cast in a context that privileges speed of production. It is not just that *presenting* a risk assessment to an audience requires rhetoric, but also that *determining* and *calculating* those risk assessments are rhetorical acts. By choosing starting assumptions with an eye to an audience's values or interests, a scientist is making a rhetorical choice, even if it is one expressed primarily in numbers rather than words.

The following section considers how Marshall determined risks rhetorically as well as scientifically, in keeping with the dominant values of the Manhattan Project. Marshall's audience included scientists and engineers who oversaw the production process at Hanford. This audience might object to an overly conservative estimate if it would slow down production, but it might also consider an insufficiently conservative estimate too hazardous to support. The most persuasive calculation of risk would be the one that seemed neither too cautious nor too risky for Marshall's audience to accept.

The second part considers how Marshall framed her risk assessments rhetorically. The intensity of wartime work on the Manhattan Project led Marshall to take risks of her own, using simplified calculation methods in order to save time. In her memos and reports, Marshall had to justify the risks she took in her own research, even when she miscalculated safety hazards at the Hanford plant. By casting her research in terms of the urgency and the discourse of technical rationality, Marshall defended these risks and argued that they were appropriate within the context of the Project.

DETERMINING RISKS

When she first toured the Hanford plant, Marshall began to worry that a phenomenon she had studied as a chemistry student might occur in the plant, with potentially dangerous results. At Hanford, plutonium was extracted from irradiated fuel elements, which were first transformed from a solid fuel into a liquid solution. Next, the plutonium in this solution was isolated from uranium and other isotopes using chemical and physical processes until it gradually became concentrated into a semisolid paste (Marceau et al. 1.31). One of the physical processes used in this procedure was centrifuging—a process that Marshall feared might lead to an unforeseen explosion. In *The Uranium People,* she describes this phenomenon: "when there is some amount of precipitate in a solution, one is better able to see it if the solution is swished around in a beaker. The precipitate then collects at the center of the swirling liquid. I was worried that when plutonium was precipitated from the solution and centrifuged to separate the liquid from the solid, the swirling motion of the liquid in the centrifuge might cause enough precipitate to collect together that a low-grade chain reaction might occur" (Libby 172). Such a chain reaction, if left uncontrolled, could lead to a potentially hazardous explosion. To ensure that no such effect occurred, scientists recommended limits on the amount of plutonium that could be processed at one time.

Marshall and her colleague, John Wheeler, took up this problem in the spring of 1945, when the production schedule at Hanford increased to accommodate the urgent demand for plutonium. Previously, in 1944, Arthur Compton had suggested that the accumulation of plutonium in storage vessels should be limited to 250 grams to ensure that no unforeseen chain reaction occurred. By 1945, as Wheeler notes in a later report, these values had to be reevaluated when scientists realized that the plant might soon produce more than 250 grams of plutonium at a time (5).

The increase in production meant that sufficient amounts of the product might accumulate in the centrifuges to sustain a chain reaction. Marshall and Wheeler wrote two reports to determine the correct amount. The first, titled *Limiting Mass,* was dated March 29, 1945, while the second, entitled *Limiting Mass: An Application of Considerations Presented in a Memorandum of the Same Title, March 29, 1945,* was dated April 9, 1945. To create an effective set of limits, Marshall and Wheeler would have to satisfy two requirements: one for maximum productivity, the other for safety. These requirements not only had to be incorporated into the data they calculated, but also had to be justified rhetorically and scientifically. Wheeler and Marshall used a discourse of technical rationality to do so.

In the first report, Marshall and Wheeler decided to give more weight to timeliness and productivity, and less to safety measures. The authors justify their approach using a distinction between "probable" and "conservative" approaches to the problem. Probable assumptions would determine limits that would make it *unlikely* that a chain reaction would occur in a given area of the plant. In contrast, conservative assumptions would produce results that would *ensure* that an unforeseen chain reaction would not occur. Although more conservative assumptions would provide increased safety against accidental chain reactions, they would also require plant operators to limit the quantities of plutonium that could be handled at one time.

Guided by the ethic of technical rationality, Marshall and Wheeler chose to use probable assumptions, which would have been persuasive to their audience of scientific and engineering officials because it ensured safety while maximizing productivity. For example, the authors chose to use a probable assumption about the number of fast neutrons that would be emitted for each slow neutron absorbed in a chain reaction. The probable assumption is that 2.0 fast neutrons are emitted for each slow neutron absorbed; the conservative assumption entails a value of 2.2 fast neutrons per slow neutron. A value of 2.0 would lead to a "safe working limit" of 360 grams of plutonium, while the conservative value of 2.2 fast neutrons would allow only 250 grams (2). In other words, the probable assumption would effectively increase the quantity of plutonium that could be treated at once from 250 grams to 360 grams, thereby increasing the overall efficiency and productivity of the operation.

Marshall and Wheeler argued that the choice of probable assumptions in this report stemmed from their awareness of the time constraints they faced. Their report states, "The chain reacting limits listed here are calculated on the most probable assumptions, rather than the most conservative ones" (2). The authors claim that an overly conservative estimate of the safe working limit could negatively affect the overall productivity of the plant. The reason for the less conservative assessment, the authors argue, has to do with the pace of production required: "To superpose upon the operating safety factor a calculational [*sic*] safety factor any more substantial than necessary might result in a serious limitation of batch size" (2). The authors argue for a probable, rather than conservative, safe working limit on the basis of time concerns, not just safety. Probable assumptions seemed more appropriate to an audience concerned with keeping production on schedule.

The report entirely excludes the broader context for this set of calculations. Marshall does not directly mention the type of explosion she feared, nor does she even allude to the potential effects of such an explosion on workers at the plant. Given the emphasis on technical rationality, the genre of the report she wrote included no provisions for such moral considerations. Nonetheless, Marshall and Wheeler certainly made ethical judgments in framing their report as a matter of "probable assumptions." The discourse of technical rationality effaced such decisions and provided no language or space in which they could be expressed.

TAKING RISKS

On April 9, 1945, Marshall and Wheeler wrote a second report on the issue of limiting mass. This time, they espoused a conservative approach to the problem. They shifted the approach not because they had changed their minds about the relative risks of safety and productivity, but because they discovered some mistakes in their first report. Marshall and Wheeler had taken a risk in determining their results, and in the second report, they had to account for that risk and reassure their audience that their calculations were valid.

Marshall and Wheeler were working under significant time pressures because the calculations they conducted were lengthy and time-consuming. During the war, these calculations were performed with a mechanical calculator, not a high-speed computer. In fact, the term *computer* referred to a person, not a machine.[8] On the Manhattan Project, women were often employed to perform these calculations; Wheeler and Marshall were assisted by Mrs. Ardis Munk, "a faculty wife," who used a "turn-the-crank mechanical calculator" (Wheeler and Ford 39). Given the protracted nature of this work, Marshall and Wheeler developed timesaving measures, using assumptions that limited the number of variables to be considered. Rhetorically, of course, it would not be savvy to call such calculations time-saving, since this might make them seem less scientifically or mathematically sound. Instead, Marshall and Wheeler called them "idealized assumptions."

For their first report, Wheeler and Marshall determined their results using idealized assumptions. However, they later realized that they had underestimated the amount of plutonium that would be required for a chain reaction to occur. Marshall and Wheeler were now confronted with a rhetorical problem. In their second report, Marshall and Wheeler had to explain why the idealized calculations were inaccurate, yet still convince their readers that the possibility of such a reaction was small and that production could continue apace.

This time, Marshall and Wheeler use conservative assumptions, rather than the probable assumptions they used in their first report. However, they also downplay the possibility that a critical reaction could occur, and then propose a different set of "idealized" calculations to determine the proper value of the safe working limit. This new set of calculations actually *increased* the amount of plutonium that could be produced at one time, even though Marshall and Wheeler portrayed these calculations as more conservative. Describing their calculations in this way would help to reassure their readers that the limits they espoused would not threaten the safety of workers at the plant. Once again, however, they argued for this set of assumptions by relying on technical rationality.

In *Limiting Mass: An Application and Consideration Presented in a Memorandum of the Same Title,* Marshall and Wheeler begin by admitting that the time-saving calculations used in the March 29 report had led them to underestimate the quantity of plutonium required to produce a chain reaction. They write: "The method of calculation was straightforward but time consuming. For this reason it was necessary to make a simplifying assumption about the spatial distribution of capture process over the plutonium-bearing part of the solution in order to obtain usable numerical results in the limited time available. A subsequent examination of this point has indicated that the simplifying assumption actually used—uniform distribution of capture processes over the solution—might be appreciably wide of the mark" (18). In other words, they had calculated the previous data using a simplified method in order to save time, but that method had skewed their results.

Next, Marshall and Wheeler describe the more conservative approach to safety used for the current report. For example, they write the following with regards to the amount of product that can be safely processed in Hanford's 221 and 224 Buildings: "Granted that we have to consider, not the most probable conditions of operation, but *the most critical con-*

ceivable conditions which are consistent with the existing process, then in the light of the three agglomeration mechanisms just discussed we can not altogether exclude the possibility that the product will be collected together in a limited region at the bottom of the centrifuge bowl" (8, emphasis added). One page later, they mention that "We calculate the critical mass on this slightly more conservative though improbable basis" (9). Using conservative assumptions helps Marshall and Wheeler to indicate that they are prioritizing safety, a rhetorical move meant to assuage potential concerns among their readers.

Yet, the authors also seek to diminish the probability that a chain reaction could happen even while they assume a more conservative or cautious approach. After admitting that the calculations from the earlier report were incorrect, the authors state that "We visualize the possibility that the product might be thrown out of solution and slowly settle as a precipitate, though this is only remotely conceivable" (12) and that "The uncertainty just mentioned is not so serious as it appears" (18). Throughout the report, they stress that a chain reaction is very unlikely to occur in the various storage tanks and centrifuges within the separations plant.

Despite assuming more conservative values, Marshall and Wheeler once again adopted an idealized set of assumptions and needed to account for them in the report. The method from the March 29 report was based on the assumption that the pieces of equipment being analyzed would be surrounded by water, which would act as a reflector for particles and accelerate a nuclear reaction. In the April 9 report, the authors note that none of the pieces of equipment under consideration was actually surrounded by water. To simplify their calculations for this report, the authors compute only the values for equipment surrounded either by no reflector (no water) or a partial reflector (some water) (18). Remarkably, the authors once again state that this tactic is preferable because it saves time: "To handle the problem of incomplete reflector by the general method of the preceding memorandum would be mathematically rather complicated" (18). Wheeler and Marshall also argue that this approach is more conservative and time-saving than the previous set of calculations: "This way of handling the problem is not only more conservative than the earlier treatment, but also easier to apply to the specialized geometries encountered in the actual process vessels" (2). Overall, they yielded a greater safe working limit than the previous set of calculations. In this way, they determined risks in a manner appropriate to the values of the Project, addressing both the need for safety and the need to keep production on a fast-paced schedule.

Ultimately, Marshall and Wheeler were successful in establishing a limiting mass that would prevent unintended nuclear chain reactions from occurring at Hanford during the war. Wheeler noted in a later memo that the limits they proposed in the first memo were approved on April 27 and 28 of 1945, although some uncertainties remained with regard to operating conditions in the plant (1). The issue of critical mass continued to occupy physicists and engineers at Hanford; Wheeler wrote two more reports on the issue in May and June of 1945.

This example demonstrates the rhetorical flexibility available to writers even within a system of technical rationality, which purportedly depends on value-neutral, mathematical equations. The scientists' rhetorical choices or starting assumptions—in this case the decision to privilege speed and efficiency—shaped the kinds of calculations they made and the arguments they made using those calculations. Marshall and Wheeler's reports demonstrate that a sense of technical rationality influenced the methods used to weigh risks, to make calculations, and to substantiate them in their reports. In the first report, the authors appealed to probable, rather than conservative, assumptions, because they would enable production to remain on schedule. In the second report, they appealed to conservative assumptions in order to reassure their readers that the results were valid, despite the mistakes Marshall and Wheeler had initially made by using idealized assumptions. Yet, both reports include an underlying premise that the proposed critical limit should be the largest quantity possible that would not significantly increase the risk of an accidental chain reaction. Adopting a different orientation toward time—a more leisurely production pace or a more conservative orientation toward safety, for instance—would have seemed inappropriate to physicists who had become accustomed to working under significant time pressure.

In fact, this same sense of technical rationality and urgency also led Marshall to take risks on a more personal level. In 1943, shortly after her marriage to fellow physicist John Marshall, she became pregnant. Although she informed the team leader, Fermi, of her pregnancy, Marshall hid her pregnancy from her co-workers: "neither he [Fermi] nor I told Walter Zinn, who was in charge of CP-2 operations, because Zinn would probably have insisted on kicking me out of the reactor building" (164). Indeed, the safety recommendations at Hanford would probably have excluded Marshall from working near the reactor. R. S. Stone, a medical doctor who oversaw health conditions at Met Lab, advised that women of childbearing age should be excluded entirely from single exposures to radiation above the daily limit of 0.1 roentgens (r), because he feared that exposure might affect women's reproductive health. (2). Presumably, even greater limits might be advised for pregnant women. Medical personnel carefully monitored radiation exposure for each worker, and Marshall had been warned of the reproductive dangers of radiation (Libby 155). However, she did not want to miss out on the exciting work of the Manhattan Project and felt that the medical warnings were overblown, describing the medical personnel as "overzealous" in monitoring radiation exposure (143).[9]

Marshall found that she could easily conceal her growing belly: "My work clothes—overalls and a blue denim jacket—concealed the bulge, and the pockets, containing side cutters, tape measure, slide rule, micrometer, pen and pencil, needle-nose pliers, small black notebook, and other such essentials, produced other bulges; as a result, my fellow scientists didn't know I was pregnant, right up to the last day" (Libby 164). It was harder to take the bus ride to and from the plant; Marshall noted that it brought her to work "barely in time to vomit before starting the day's work" (165). Yet Marshall worked right up until the second-to-last day before giving birth, and even then, left only because she was hospitalized for high blood pressure (164). She returned to work one week after giving birth to her son. Marshall writes proudly that her scientific work was not disrupted: "I had been measuring the neutron absorption of the new element plutonium [. . .] We went on with the measurement as if there had been little, if any break in the work" (165). Marshall's son was born healthy, and her mother agreed to take care of him so that she could continue her work.

In a sense, Marshall performed the ethic of technical rationality expected of scientists involved with the Project. By prioritizing the success of the Project over her own physical health, Marshall demonstrated her dedication to the Project. In this way, she also showed that she could perform the work of a scientist as well as a man could.[10] Although in her published reports, Marshall did not identify as a feminist, her *actions* might represent an enactment of a liberal feminist position. Such a stance assumes women's equality with men, sometimes without recognizing gender differences. As Schiebinger writes, such a position "all too often *requires* that women be like men—culturally, or even biologically (as when expectations are that working women need not take off time to have children)" (*The Mind Has No Sex?* 275).[11] By working the same hours, executing the same types of tasks as male physicists, and accepting risks to her own health, Marshall performed the technical rationality she considered appropriate to a (male) research physicist, and she justified this performance based on the urgency of the Manhattan Project. Her actions literally embody gender neutrality and technical rationality, seeking to overcome any suggestion that, as a woman, she could devote less time to her scientific career, or that working in a radioactive environment presented a risk to her fetus.

Marshall's actions also enabled her to perform the role of the scientific hero who endures bodily danger for the benefit of science. (Marie Curie, who died from aplastic anemia caused by radiation exposure, is another famous example.) Marshall mentions that, prior to becoming pregnant, she willingly exposed herself to a high dose of radiation in order to package and transport a small canister containing a sample of radium-beryllium to New York: "We also both sustained a large dose of radiation from gamma rays emitted by the source during the 2 hours or so that we were soldering it. I received about 200 roentgens, and my white-cell count dropped to half its usual level. However, the source didn't leak; the job was correctly done, and it had to be done. The only outward symptom of the exposure to the radiation was that I felt a bit tired for a while. The doctors of the Metallurgical Laboratory gave me a lecture on their belief that a woman has only so many egg cells, and if these are destroyed, that's it" (Libby 155). Marshall may have simply been swept up in the dominant sense of urgency that shaped research at the Manhattan Project. Performing technical rationality enabled her to demonstrate her equality as a scientist and her seriousness about furthering her career. Yet, this example shows how the discourses of urgency and technical rationality pervaded the culture of the Manhattan Project, shaping bodies and behaviors as well as texts.

Katharine Way: Managing Safety

Like Marshall, Way researched safety hazards at the Hanford Plant, but Way seemed more concerned about safety hazards. Yet, the tensions between three different organizational structures—military, scientific, and technical—meant that Way had to address her concerns not only to fellow scientists, but to engineers and military personnel. Technical rationality was embedded in these organizations and the genres that supported them in ways that limited scientists' attempts to fully evaluate the ethical, safety, and environmental impacts of their research on radioactive substances. As a junior woman, Way found it difficult to voice her concerns about safety within this organizational context.

HANDLING ORGANIZATIONAL STRUCTURES

The Hanford plutonium plant united military, scientific, and industrial personnel, creating tensions between three different types of institutional cultures. While in the earliest stages of the Project, scientists were in control, military personnel also became involved once General Groves assumed command of the Project in 1942. Scientists preferred a looser, more horizontal organizational structure to the hierarchical, military model Groves imposed. In his autobiography, for example, Wheeler writes that "The administrative structure at the Met Lab was loose and informal. I may have been assigned officially to Wigner's group, but I did not 'report' to him" (Wheeler and Ford 39). Wheeler notes that Leona Marshall was technically Fermi's assistant (55), but Marshall writes that the two had a close working relationship. Marshall did assist Fermi with calculations, but the two also coauthored a paper together after the war, and during the war, went swimming in Lake Michigan every evening in the late summer and early fall (Libby 19, 7).

In contrast to this informal structure, Groves imposed "command channels" that, while somewhat more flexible than traditional military structures, assumed a hierarchical organization (Groves xv). At Hanford, a third type of structure emerged when E.I. DuPont took over the management and construction of the plants. Scientists sometimes found it difficult to work with DuPont's engineers, who had no experience in this type of production, and who sometimes questioned the scientists' design plans. These different organizational cultures created tensions within the Project, especially given the overall emphasis on timely completion. Because the design and development of the plant occurred concurrently, there was even more overlap between these three corporate cultures than there might have been otherwise.

The values guiding the Project also created tensions. The design and operation of reactors followed three main rules. General Groves writes that these included: "1, safety first against both known and unknown hazards; 2, certainty of operation—every possible chance of failure was guarded against; and 3, the utmost saving of time in achieving full production" (83). Groves's rules prioritize safety, certainty, and timeliness, three (sometimes conflicting) factors that guided the design and development of the Hanford Plant. Scientists and engineers had to ensure that the Project proceeded on schedule without endangering personnel or people living in the areas surrounding the plants.

In contemporary terms, one might consider safety, certainty, timeliness, and technical rationality among the values that characterized the "corporate culture" of the Hanford Plant. Linda Driskill writes that corporate culture and organizational structures play a fundamental role in shaping how writers understand rhetorical situations (65). In turn, writers' understandings of the rhetorical situation shape rhetorical choices, including "which events or perceptions count as facts, which concepts apply to these facts, and which assumptions are used to evaluate them. The definition of situation determines which words are chosen as appropriate to the subject, which roles are available, which range of actions is appropriate; and with whom one is to communicate and how" (65). As Driskill suggests, corporate structures and cultures play an important role in shaping scientific and technical knowledge—what counts as facts and how those facts are to be interpreted. I would add that corporate cultures also determine a collective sense of what kinds of issues and questions are deemed timely and appropriate. In her work on reactor design, Way encountered just such a disparity in terms of the timeliness of her claims. While she found issues of radiation safety important and timely, others granted those issues less exigence. Due to the hierarchical culture imposed on the Manhattan Project, as well, Way found it difficult to question the results she received from a male engineer, one who valued technical rationality above all else.

CHALLENGING EXPEDIENCY

In July, August, and September of 1944, a discussion occurred between Katharine Way and three other individuals: Samuel Allison, Charles Wende, and, indirectly, Eugene Wigner. Allison, Wigner, and Way all worked from the Met Lab in Chicago. Allison was director of the Metallurgical Laboratory, while Wigner led research on the design of the Hanford plant. Presumably, Wende was an engineer who worked for DuPont.[12] (Unfortunately, he does not provide his title in his letters.) Way found her results questioned and her facts refuted. In contrast, when Way questioned Wende's data, Wende countered her challenge and upheld his facts as the correct ones. This exchange illustrates how rhetorical and scientific cultures can limit individual scientists' ability to raise ethical questions about their research.

The exchange began, so far as archival records indicate, on July 8, 1944, when Way wrote a memorandum to Wende (shown in Figure 3.1). She writes: "Dear Charlie: In one of your recent weekly reports, N-1258 to us, you say that under normal operating conditions a man may spend three minutes daily in the discharge area. I am unable to check this result and wondered if you would be kind enough to go over the following figures to see where the discrepancy is. Perhaps this discrepancy is the

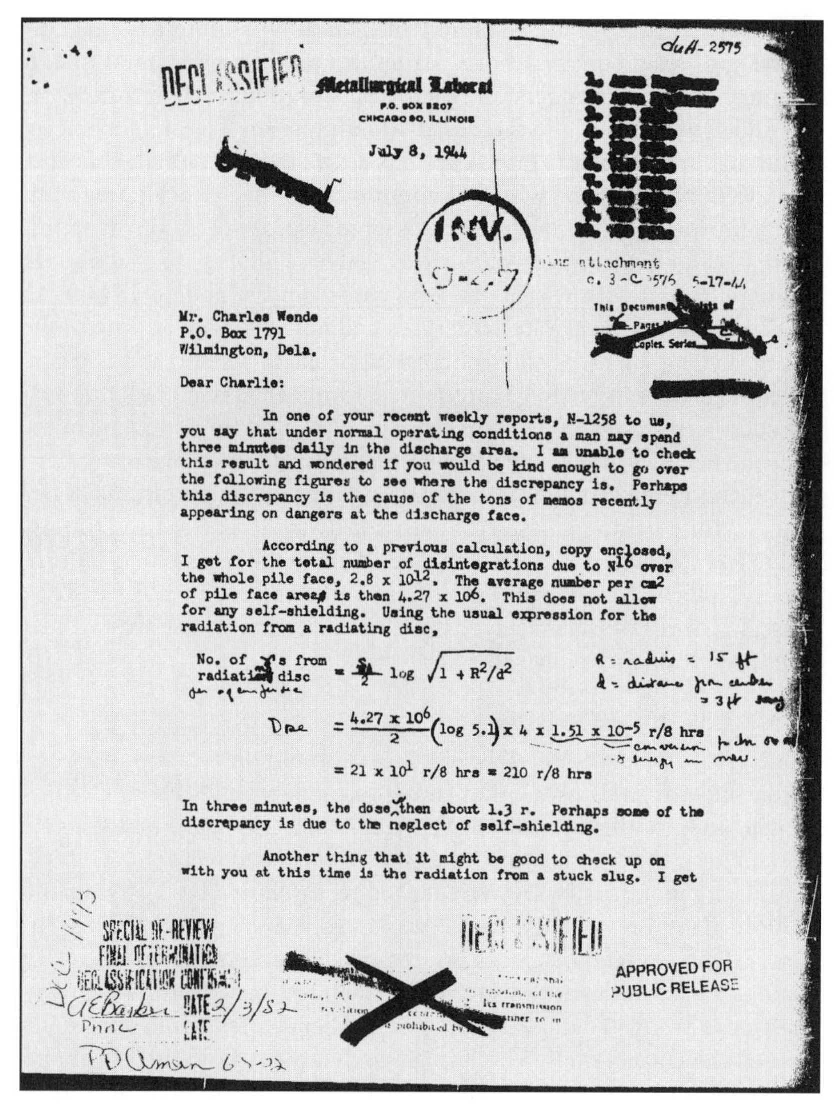

DECLASSIFIED

Metallurgical Laborat

P.O. BOX 5207
CHICAGO 80, ILLINOIS

July 8, 1944

Mr. Charles Wende
P.O. Box 1791
Wilmington, Dela.

Dear Charlie:

In one of your recent weekly reports, N-1258 to us, you say that under normal operating conditions a man may spend three minutes daily in the discharge area. I am unable to check this result and wondered if you would be kind enough to go over the following figures to see where the discrepancy is. Perhaps this discrepancy is the cause of the tons of memos recently appearing on dangers at the discharge face.

According to a previous calculation, copy enclosed, I get for the total number of disintegrations due to N^{16} over the whole pile face, 2.8×10^{12}. The average number per cm^2 of pile face area is then 4.27×10^6. This does not allow for any self-shielding. Using the usual expression for the radiation from a radiating disc,

No. of γ's from radiating disc $= \frac{S_A}{2} \log \sqrt{1 + R^2/d^2}$

R = radius = 15 ft
d = distance from center = 3 ft

$$\text{Dose} = \frac{4.27 \times 10^6}{2} (\log 5.1) \times 4 \times 1.51 \times 10^{-5} \text{ r/8 hrs}$$

$$= 21 \times 10^1 \text{ r/8 hrs} = 210 \text{ r/8 hrs}$$

In three minutes, the dose is then about 1.3 r. Perhaps some of the discrepancy is due to the neglect of self-shielding.

Another thing that it might be good to check up on with you at this time is the radiation from a stuck slug. I get

SPECIAL RE-REVIEW
FINAL DETERMINATION
DECLASSIFICATION CONFIRMED
DATE 2/3/82

DECLASSIFIED

APPROVED FOR
PUBLIC RELEASE

Figure 3.1. Katharine Way's memo to Charles Wende, 8 July 1944; image courtesy of the Hanford-Battelle Public Reading Room, Richland, Washington.

cause of the tons of memos recently appearing on dangers at the discharge face" (Letter to Charles Wende, 8 July 1944). According to Way's own calculations, the dose of radiation received in three minutes at the discharge face would be "about 1.3 r" (1). This value would be much higher than the daily permissible dose of 0.1 r per eight-hour day, so it was important to Way to verify the result. If Wende's result was correct, workers would be allowed to work in that area for three minutes. But if Way's result was correct, they would be barred from that area altogether.

Way phrases her challenge to Wende gently and indirectly, using rhetorical strategies that Priscilla S. Rogers and Song Mei Lee-Wong have identified as hallmarks of politeness. Rogers and Lee-Wong write that politeness strategies "center on showing sensitivity to the face needs of the receiver of the communication by mitigating face-threatening acts, such as requesting, criticizing, or refusing" (382). These strategies, they point out, are especially important for subordinates writing to those in a position of power—a position women have historically been especially likely to occupy.

Way uses such "politeness strategies" as "deference," which is used to show respect for the reader's knowledge and expertise and respect for the reader's own responsibilities and time constraints (Rogers and Lee-Wong 400). In her memo, by writing "I am unable to check this result," Way places the burden of proof not on Wende, but on herself. Rather than requesting directly that Wende review his figures, she writes that she "wondered if you would be kind enough to go over the following figures to see where the discrepancy is." She also alludes to situational factors, such as the "tons of memos recently appearing" due in part, to the frenzied pace of research on the Hanford design. Rhetorically, Way's memo seems to employ deference quite effectively to avoid the implication that Wende's results were incorrect.

In her letter, Way also suggests some possible reasons for the discrepancy between her result and Wende's. She hypothesizes: "Perhaps some of the discrepancy is due to the neglect of self-shielding" (1). Way also invokes, somewhat subtly, the authority of Samuel Allison, a senior physicist at Met Lab: "I am writing to you with the knowledge of Mr. Allison because it seems as though we may really be using different fundamental figures in questions of the discharge area" (2). Thus, Way tries to maintain goodwill by using deference and by phrasing her question in a nonthreatening manner. She appeals to outside factors to displace blame for the discrepancy, and appeals to her superior, Allison, in order to underscore the need to settle the discrepancy. Way seemed well aware of politeness conventions typical of scientific correspondence.

Nonetheless, Way's plea was not well received. In his response, dated August 1, 1944, Wende states that the discrepancy is no longer important: "No change in the operation of the pile will result from an upward revision of estimates on the hazard in this area, since the lowest estimate was already too high to permit access during operation" (Letter to

Katharine Way, 1 August 1944). This is a curious statement, considering that Wende's previous report, as Way quoted it, stated that workers could work in that area for three minutes.

Next, Wende remarks that he will explain the discrepancy between his results and Way's anyway, "To keep the record straight" (1). He suggests that Way's method of calculation was inappropriate: "It strikes me that the use of the circular disc formula in a case of this type may be *questionable.* A more *reasonable* upper limit might be obtained if the flux out of the source is taken equal to half the number of disintegrations per cm^2 sec" (2, emphasis added). Wende implies that the discrepancy is due not to mathematical error, but to a questionable and unreasonable method on Way's part.

While Way phrased her questions neutrally and declined to assert that her own estimates were preferable, Wende did not hesitate to claim that his estimates were correct: "The remaining discrepancy is due mainly to the assignment of pile size and power. In any case, I think that a figure of the order of 100–130 r/8 hr looks like an upper limit for the smaller pile at high power, and that in practice something in the range of 20–40 r/8 hr. will be more probable" (2). Overall, then, while this exchange began with Way questioning Wende's results, Wende was able to turn it around, suggesting that Way's results were derived from "questionable" methods, yielding "unreasonable" results. While Way seemed to err on the side of deference, Wende's letter erred more on the side of confidence—he felt free to assert authority in his letter and to claim his results were more accurate and reasonable.

A similar exchange occurred the same month, again regarding a discrepancy in calculations. In this case, Way writes to Allison and implies that he might take up the issue with the DuPont managers. Once again, however, Way's colleagues challenged her calculations. This time, the exchange began with a memo Way wrote to Allison on August 11, 1944 about a radiation hazard in a different area of the plant, the control rod room. In this case, Way suspected that radioactive activity in the water that cooled the control rods would make the room dangerous during pile operation (memorandum to Samuel Allison, 11 August 1944). This danger would be increased if there were any "unnecessary hose on the reels" (1). Way explains that the room needed only about forty feet of hose, but that fifty feet were planned. The additional ten feet of hose would increase the amount of radiation in the room, so it was important that the length of the hose be reduced as much as possible. She writes: "It is my impression that the duPont [*sic*] Company is aware of the danger here but perhaps a quantitative estimate of its magnitude, based on the recent exact account of Mr. Jesse's experiment (CE-1632) will be of help to all concerned" (1). Way includes Mr. Jesse's experimental results in the body of the letter, concluding that the dose in the control rod room would be 0.34 r/hr. Safety measures would need to be enacted to minimize exposure to radiation. The letter Allison wrote to Wende is missing, but in his reply to Allison, dated September 1, 1944, Wende indicates that Allison had written to Wende on August 16 and enclosed "Miss Way's calculations of the radiation danger from control rod cooling water" (Letter to Samuel Allison, 1 September 1944). Wende neglects to mention that the calculations Way discusses were done by Mr. Jesse, whom Way mentions twice in her letter, not Way herself. Once again, Way was blamed for faulty calculations, even though the calculations were not her own.

In his response, Wende downplays the risk Way had highlighted: "It is planned to monitor the apparatus room carefully during initial operation at low power and to take whatever course is found necessary to minimize the danger. If a gamma hazard is found, it may be necessary to limit access to this area. If a beta hazard is found, additional shielding can easily be installed" (Letter to Samuel Allison, 1 September 1944). At this point, Wende once again questions Way's results as being too high: "The estimate of the gamma hazard given by Miss Way may be high by a factor of five or so" (1). He provides various reasons why a lower estimate could be used, mostly based on additional variables in the design of the plant, such as the use of aluminum instead of heavy metal slugs and the position of the control rod (1). Wende notes also that radiation from beta dosage "can be remedied by placing heavier sheet metal covers over the hose lines; and this is what Operations plans to do *if this danger appears*" (1, emphasis added). Even though Way had invoked Jesse's calculations instead of her own, Wende described the data as "Miss Way's calculations" and once again deemed them inappropriate.

As a response, Allison wrote a letter to a W. O. Simon of DuPont, with the subject line "Attention: Mr. C. W. J. Wende." He states that he referred Wende's September 1 letter to Way and also Eugene Wigner, an even more senior physicist, for their comments. Allison argues that a slightly high estimate of the radiation hazard would help to ensure the safety of workers at the plant. He writes that Way and Wigner "are agreed that Mr. Wende's statement is correct that the estimate of the gamma hazard previously given by Miss Way may be high by a factor of 5 [. . .] However, both Miss Way and Mr. Wigner feel that there should be some safety factor in the shielding estimates" (Letter to W.O. Simon, 22 September 1944). At this point, the communication chain drops off, at least from the available archival materials, so it is unclear what happened next. Nonetheless, it is significant that Way's colleagues second-guessed or refuted her calculations on two different occasions. Way tried two approaches with Wende, one invoking the horizontal communication structure, and the other invoking a more hierarchical structure. Yet the results were the same.

One interpretation, of course, would be that Way lacked care and accuracy in her scientific calculations. Another interpretation might be that this interchange simply reflected differences in status. As Rogers and Lee-Wong suggest, professional communication often requires subordinate writers to balance deference with confidence. An overly deferent approach might prevent a writer from coming across as competent. While subordinates must maintain a tone of respect in their writing, then, they must also be willing to demonstrate their ability and expertise (396). Perhaps Way erred too much on the side of deference and failed to show enough confidence in her memos. At any rate, the discourse of technical rationality required Way to phrase her concerns in strictly quantitative terms, and the conventions of politeness typical in the genre of the technical memo prevented her from making an emphatic case for the safety measures she felt necessary.

The Discourse of Technical Rationality

Overall, a network of conflicting factors within the organizational culture of the Manhattan Project, including status, time, genre, and institutional constraints, limited potential discussions of safety and ethics. Because scientific research on safety hazards proceeded simultaneously with the design of the plant, scientists had to deal with a continually shifting set of variables, such as the length of hose in the control room or the amount of shielding around a radiation source. As a new and temporary organization, the Manhattan Project did not have a well-established organizational structure or culture, but more of an ad hoc culture that arose out of three distinct groups, each with its own traditions and concerns. Scientists like Way had to negotiate this uneven set of tensions and to seek a place from which to write persuasively. Further, both Marshall and Way used technical rationality as the only discourse with which to discuss safety issues. Within the genre of the technical report adopted on the Project, no other vocabulary was available with which to raise ethical or safety concerns.

Unfortunately, for a variety of reasons, including her cautious approach to risk, her subordinate status as a junior scientist, the collective sense of technical rationality and perhaps her gender, Way's calculations seemed "inappropriate" and untimely. The perceived urgency of the Manhattan Project provided a justification to discredit estimates of risk that placed too much emphasis on safety and too little emphasis on timeliness. In contrast, Marshall's calculations were accepted in part because she was willing to assume a less conservative approach toward risk. Ultimately, given the time constraints that shaped the Manhattan Project, the *genre* of the scientific report that emerged seemed to be one that valued riskier calculations over more conservative ones. Within these genres, there was neither an appropriate space nor vocabulary with which to raise ethical issues.

The time pressures that shaped the Manhattan Project seemed to gain energy as time went on. Spurred on by a collective sense of urgency, scientists and engineers completed an enormous undertaking with limited knowledge and even more limited time. Yet, as the Project gained momentum, scientists themselves had less control over the atomic bomb and how it would be used. On July 17, 1945, scientists at the Met Lab wrote a petition urging the government to exhibit the bombs in an unpopulated area rather than to drop them without warning on Japan. Way signed the resulting document, along with 69 other scientists from Met Lab and Clinton Engineer Works; Marshall did not. The petition mentioned specifically that the decision to use the atomic bomb would create a chain of devastating effects stretching far into the future: "The development of atomic power will provide the nations with new means of destruction. The atomic bombs at our disposal represent only the first step in this direction, and there is almost no limit to the destructive power which will become available in the course of their future development. Thus a nation which sets the precedent of using these newly liberated forces of nature for purposes of destruction may have to bear the responsibility of opening the door to an era of devastation on an unimaginable scale" (Szilard *n.p.*). However, the petition did not succeed in persuading the president to forestall lethal use of the bomb. The wartime context called for action—not discussion.

As the government continued to pour money into the Manhattan Project, the government also became increasingly invested in the eventual use of the bomb, which was to end the war as quickly as possible in order to save American lives. Ultimately, urgency and technical rationality made long-term considerations seem unimportant for many focused on the outcome of the war. Those who wished to argue for higher safety standards could do so using only the technical vocabulary privileged within the Manhattan Project. This restriction has become commonplace in contemporary discourses about technology as well; as Langdon Winner laments, "Unless one can demonstrate conclusively that a particular technical practice will generate some physically evident catastrophe—cancer, birth defects, destruction of the ozone layer, or some other—one might as well remain silent" (51). This was the unfortunate lesson Way seems to have learned through her unsuccessful exchanges with Wende.

However, wartime decisions made during the war have had far-reaching effects on human and animal lives, even while they purportedly saved the lives of thousands of American soldiers. The urgency that justified nuclear developments during the war years has begotten new problems. For instance, time pressures from General Groves led DuPont to accelerate production of plutonium at Hanford, running reactors above their rated power level and shortening the cooling period for irradiated fuel elements to as little as a few weeks, as opposed to the longer periods recommended for safety reasons (Marceau et al. 1.27). In a 1944 report, for example, Way noted that this waiting period was normally taken to be sixty days, but that tight deadlines might cause that period to be shortened (Way, *Time Interval* 1).

The effects of these and other time-saving measures became apparent only after the war. Scientists later found that Hanford's storage containers for nuclear waste, built rapidly during the war, had leaked half a million gallons of radioactive liquid into the ground. In *The Uranium People,* Marshall attributes this leak to the urgency of war: "We were in such a hurry back in 1943 to construct and operate Hanford in wartime, with its concomitant shortages, that the stainless-steel storage tanks could not wait for stainless-steel pipes and, instead, were connected with cast-iron pipes. Consequently, an electric battery was set up at the interface between the different metals, with the result that corrosion began and continued for some 33 years until leakage set in" (Libby 190). Given the tendency to focus on speed of production during the war, such concerns seemed untimely and out of step with the prevailing ethic of technical rationality. As it turns out, storing nuclear waste requires systems that will last for 10,000 years into the future—the amount of time it takes for nuclear waste to decay to a point where its radiation levels would be "acceptable." Scientists knew this during World War II, but many assumed that they could find a solution to this problem in the future, perhaps by reusing nuclear waste in some way.

Although the Manhattan Project included a number of scientists who studied the health effects of radiation—including the now-infamous Human Radiation Experiments—they still did not know (or claimed not to know) whether the genetic effects of even minimal exposure to radiation might show up many years after exposure. Groves summarizes that "no one had any reasonable idea of what the hazards might be or the likelihood of their occurring [. . .] no one could predict the duration of the effects of the hazard, or [. . .] even when the effects might first appear [. . .] [and] no one could possibly predict the extent of the damage if a major catastrophe occurred" (57). Nonetheless, in Groves's terms, "time was of transcendent importance," and pursuing the Project took precedence over these concerns (57). So scientists and engineers did their best to limit exposure to radiation at the time.

Further, the Department of Energy admitted in 1990 that emissions of iodine 131 from Hanford's plutonium production plants between 1942 and 1946 may have caused thyroid cancers or other diseases in people living in neighboring areas (Karen Steele *n.p.*). According to the United States Environmental Protection Agency, radiation, depending on the dosage received, can increase one's risk of cancer, as well as the risk that one's children will suffer from genetic defects or mental retardation (Environmental Protection Agency *n.p.*). Although the possibilities of these effects were recognized during this period, they were not considered threatening enough to stop such research altogether.

Ultimately, individual interpretations of the rhetorical situation seem to have shaped Marshall's and Way's own assessments of the impacts of the nuclear bomb. Marshall's postwar writings suggest that she supported the use of the atomic bomb. In *The Uranium People,* she writes that "My brother was part of that operation [landing of Marines at Okinawa] and, without doubt, he would have been among the first to land on Japan. I have no doubt that the bombs saved his life. My brother-in-law was an officer on the mine-sweeper scheduled first into Sasebo Harbor in the planned November invasion. He would surely have died" (Libby 244). Marshall seems to have identified with the purposes of the Manhattan Project. For her, the bomb was justified because of the temporal pressure to end the war and save human lives.

After the war, Marshall advocated for peaceful uses of atomic energy. The final chapter of *The Uranium People* takes up and refutes objections that nuclear reactors are unsafe and that they create problems for storage of fission products. Appealing to scientific progress, Marshall writes that storage of fission products would not be a problem in the future because those products would be used for new inventions (334). Marshall also imagined a positive future outcome of wartime research, namely, that peacetime application of atomic research would provide a safe, plentiful source of energy (326).

In contrast, Way seems to have been more cautious about the role of nuclear energy in the postwar world. In addition to signing the Met Lab petition mentioned earlier, after the war she joined efforts to pressure the American government to ensure civilian control of atomic energy. She coedited with Dexter Masters the postwar collection *One World or None* (1946), which includes reflections from eminent scientists on the future of atomic energy and the need for international civilian control to ensure its proper application. Masters writes that "Many scientists in the wake of the Japanese bombings were concerned by what they saw as a general lack of understanding of the bomb's threat. But to the best of my knowledge, it was Kate Way who brought the concern to focus" (qtd. in Martin et al. 574). Although she continued her research career after the war, Way also pursued issues of social justice, taking part in the civil rights movement of the early 1960s and later advocating for senior citizens in North Carolina (Martin et al. 575).

This chapter demonstrates that technical rationality can be both enabling and disabling, and that its effects differ across axes of time, status, gender, and genre. On the one hand, the sense of intensity and expediency helped scientists and engineers to complete an enormous undertaking with limited knowledge and time. On the other hand, the intense focus on timing that emerged during the Manhattan Project produced what Kenneth Burke has called "trained incapacity," or "the state of affairs whereby one's very abilities can function as blindnesses" (3). Many Manhattan Project scientists became so adept at dealing with urgency and technical rationality that they were blinded to the broader ethical implications of their research. Some scientists protested the use of the bomb, but there was little time for reasoned debate and even less chance for those who wished to forestall its use. Although technical rationality led Marshall to downplay safety issues, conventions of politeness in technical memos constrained Way's attempts to raise such issues. The ethic of technical rationality employed on the Manhattan Project provided no rhetorical space for discussing issues of human health, environmental safety, and ultimately, the use of the bomb itself.

4. *Women Nutritionists on the National Research Council*

> Wars are won or lost according to the health, courage, and morale of whole populations and their ability to exert themselves to the utmost, and this is particularly true in modern total warfare.
>
> —Recommendations of the National Nutrition Conference to the President of the United States, 1941

In the fall of 1940, as America readied its defenses for war, the National Research Council's Food and Nutrition Committee moved to position nutrition within the country's defense program. They aimed to promote developments in nutritional science, and to advise government agencies on the nutritional needs of the armed forces, the civilian population, and overseas populations needing food relief. To provide a scientific grounding for this work, they needed a complete set of dietary standards to determine the quantities of each nutrient necessary to maintain optimum health. The committee chose three female scientists to draw up a list of the minimum dietary requirements. (Overall, the group included twenty-five men and four women.)

As a leader in the field of nutrition, Lydia J. Roberts was an obvious choice to lead this initiative. As she recounts the story of this major wartime project, "Dr. [Russell] Wilder appointed a group of three—Dr. Helen Mitchell, Dr. Hazel Stiebeling, and myself—to prepare such a set of standards during the evening and be ready to present them to the group the next morning! We three spent the evening threshing over the problem (while the men, we felt sure, were out seeing the town). The result was, of course, that the only report we could bring in the next morning was that it couldn't be done, that the evidence was too scanty and too conflicting" ("Beginnings" 903). That night, Roberts and her committee drew up a preliminary list of standards, but urged the committee to conduct more research on a set of standards that would have greater scientific credibility. This would ensure that the standards would gain the approval of those who would be using them, including home economists, nutritionists, dietitians, journalists, government agencies, medical associations, welfare associations, commercial food groups, public health departments, and extension agents.

In her reports, presentations, and interpersonal communications with members of the committee, Roberts employed what she termed the "democratic approach" as a rhetorical resource. This approach, which entailed involving multiple stakeholders in decision-making processes, helped Roberts gain consensus from the various experts with whom she worked, while constructing for herself a persona as a fair-minded chairperson. Once the RDAs were developed, however, Roberts used the democratic approach to shore up the authority of the National Research Council (NRC), limiting access to the genres of power and authority to research experts.

At the same time, famed anthropologist Margaret Mead joined the National Research Council's Committee on Food Habits, where she led efforts to study food patterns in their cultural context. Mead instituted an anthropological approach to studying food habits, one that led to a series of reports on customs among various American subcultures. Yet, Mead's work also served to shore up the authority of nutrition workers, especially anthropologists who sought to intervene in a field of inquiry previously dominated by nutritionists and home economists.

Both Roberts's and Mead's wartime contributions have been overlooked in our histories of scientific rhetoric, perhaps because they wrote in the context of the National Academy of Science (NAS) committees with which they worked, rather than as individuals. In these reports, Roberts and Mead employ a "discourse of expertise"—that is, they employ abstract, scientific terminology, and appeal directly to scientific authority. Using this discourse prevented everyday citizens from contributing to nutrition outreach. They did not fully enact the participatory ideals feminists might hope for in a scientific discipline in which the proportion of women was relatively high. Roberts's and Mead's examples suggest that women who make it to the upper echelons of a scientific community often have a tough time enacting changes that encourage greater access to science for other women. One reason for this may be that they have internalized dominant discourses of science, such as the discourse of expertise, which make it difficult to appreciably alter scientific culture.

The Gendered Cultures of Nutrition and Home Economics

Although her expertise was in nutrition, Roberts's academic appointment was in home economics, a discipline founded in the early part of the twentieth century as an outlet for women who were interested in science, but prevented from pursuing a career in other scientific fields. As a female-dominated discipline, home economics held a tenuous position within most universities, lacking prestige, funding, and power (Nerad 1). Perhaps for this reason, the greater number of women in home economics and nutrition did not radically disrupt the hierarchical form of other scientific disciplines (Kohlstedt 5). Instead, it embraced a discourse of expertise and extended it outward. Home economists at universities and colleges trained generations of secondary and grade school teachers, extension workers, dietitians, social workers, and others, all of whom could claim expertise over the health and nutrition of the American housewife and her family.

The genres privileged in nutrition and home economics differed somewhat from other scientific fields because writing for public audiences was considered integral to the discipline, not peripheral. From its inception, home economists had considered writing for the public to be an important part of their mission.[1] In addition to writing scientific articles and reports, nutritionists and home economists wrote genres directed to public audiences: government bulletins, textbooks, pamphlets, radio spots, editorials for newspapers and magazines, and so on. Those with the most status in the field were not excluded from these popular genres; indeed, experts like Roberts had plenty of experience writing for public audiences. American housewives, her main audience, had little input into the content of nutritional advice; instead, they were positioned as passive consumers of information.

This institutional arrangement bolstered the expertise of home economists. In her history of the discipline, Marjorie M. Brown writes that home economists tended to have a low opinion of the average American family, and sought to impose "rules for right living" through education, advice, and legislation (324). This perspective, Brown writes, leads to a technocratic system that "undemocratically removes citizens from their political role in forming public policy and even in having their own needs recognized and given equal consideration (to the needs of others) in policymaking" (324). In this way, home economists supported a technocratic form of government that placed scientific experts in the service of the bureaucratic elite, not the publics they purported to serve.

As members of the National Research Council, Roberts and her colleagues were well placed within this technocratic system. Since its inception in 1916, the NRC has positioned itself solidly as an expert mediator between public audiences and the government. While it is a privately funded entity, the NRC and its parent organization, the National Academy of Sciences, have traditionally received government grants and contracts to further their work. That work, specifically, has been to advise government experts on scientific and technical issues, including nutrition (Hilgartner 21). Rather than addressing their reports to lay audiences, the NRC directed its efforts primarily toward other experts (i.e., government officials). They issued advisories and bulletins to be read primarily by those audiences, and only secondarily translated those materials into material useful for the public.

Roberts herself was chosen for the NRC's Committee on Food and Nutrition by virtue of her expertise and credibility in the field of nutrition. Roberts was full professor in the Department of Home Economics at the University of Chicago. She enjoyed a remarkable career, spanning the six decades between 1919 and 1965. She began her early career as a schoolteacher, during which time she developed an interest in children's nutrition. In 1915, she enrolled in the Home Economics program at the University of Chicago, earning a Bachelor of Science degree in 1917 and a master's degree the following year. Roberts started working as an assistant professor at the University of Chicago in 1919, where she embarked on clinical research on children's nutrition. This research culminated in a book, *Nutrition Work with Children* (1927), the first textbook to address the special nutritional needs of children, which earned Roberts her Ph.D., a reputation as a leader in the field, and, in 1930, a full professorship (Harper 3700). The expertise of the NRC, along with Roberts's own standing in the field, enabled her to promote the RDAs effectively among scientists, government and military officials, as well as the broader American public.

Nutritionists had already claimed a role for their research in the defense effort. The war offered an opportunity for nutritionists and home economists to increase their expertise and visibility in the public eye, using scientific appeals along with appeals to American values of patriotism, duty, and national defense. Even before the United States entered the war in 1941, the government began to bolster the nation's defenses through intense industrial and military preparation, which President

Roosevelt deemed "total defense" (10). The notion of total defense meant expanding industrial factories to produce uniforms, guns, tanks, ships, warplanes, and ammunitions, and increasing military recruitment and training. However, total defense was also broadened to include any factor that would strengthen the nation's preparation for war. As the United States geared up to produce weapons technologies, dietitians and home economists offered a different technology that could also help to win the war: nutrition.

By 1940, nutritionists and dietitians had already begun to articulate how women in their profession could contribute to total defense, using the phrase "second line of defense" as a rhetorical trope. While the "first line of defense," the military, was made up predominantly of men, the notion of the second line of defense justified and encouraged women's participation in the war effort. For instance, a 1940 editorial in the *Journal of the American Dietetic Association* (*JADA*) proclaimed: "America, finally awakened, is preparing for defense. In the wake of the first line of defense, the armed forces, must come that second line, those who heal and save lives and those who conserve the country's food supply" ("Editorial: Preparedness" 683). Nutritionists were able to position their research at the vanguard of this second line of defense by connecting nutrition to a range of military, economic, governmental, patriotic, scientific, and even eugenic discourses that circulated during the early years of the war.

Media reports helped to create exigence for the second line of defense as the nation stepped up its preparations for war. In the early stages of military conscription, the Army reported that the rate of rejections due to physical deficiencies was high—about 15 percent of the 14,500 men who were conscripted in the first round of summons in 1940 ("Draft Rejections Surprisingly High" 48). Experts argued that these rejection rates could be reduced through better nutrition and health, since many of the rejected draftees had bad teeth or were over- or underweight (Leone 1283). Nutritionists seized this opportunity to argue that they could help solve this problem through research and educational efforts. Mitchell, then principal nutritionist for the Federal Security Agency, wrote in 1941 that one-third of all draftees rejected for service demonstrated some form of malnutrition (537) and argued that it was the responsibility of nutrition researchers to translate their knowledge into practical recommendations worded in ways the public could understand, not the "mumbo jumbo of the laboratory" (538). Nutritionists were not just scientific experts, but were also expert popularizers, capable of making scientific knowledge accessible for the public.

Their jurisdiction stretched beyond the military recruit to include the American population in general. Experts feared that Americans were malnourished and would be ill prepared to undertake the physical and mental labor required for wartime defense. Citing one of Stiebeling's nutrition studies, M. L. Wilson, head of the national program of nutrition under the NRC, noted that as many as 45 million Americans were poorly nourished in 1940, due either to low income, poor food habits, or limited knowledge of nutrition (16). In a speech at the Opening General Session for the American Dietetic Association in 1940, Wilson argued that nutritionists and dietitians needed "to raise the nutritional status of the nation as rapidly as possible and to as high a level as possible [. . .] If this nation should be forced into war, then every man, woman, and child in the United States should be prepared by being in the best physical and mental condition that the science of nutrition can develop and maintain" (19–20). Good nutrition became crucial to total defense because it helped to maintain a healthy civilian population, one capable of working long hours in the factories springing up to boost production of war-related goods. By encouraging Americans to follow scientific principles when choosing their meals, and by providing guidelines for dietitians working in defense plants, experts hoped to bolster the health of defense workers. The war seemed to offer a unique opportunity to foreground the importance of nutrition, to make improvements in American's eating habits, and, simultaneously, to increase the expertise of the nutritionist. Both Roberts and Mead played a major role in this effort.

Lydia J. Roberts: Establishing Daily Requirements

Despite the obvious need for a set of nutritional standards, developing and implementing the RDAs was complicated by issues of expertise, both for the woman in charge of developing them, and for the field of nutrition as a whole. The following sections show, first, how Roberts enacted her role as chair of the Committee on Dietary Allowances in internal negotiations, using the democratic approach as a rhetorical resource to boost her own ethos. First, I examine minutes, reports, and transcripts of meetings of the National Research Council's Food and Nutrition Board and of the Committee on Dietary Allowances. Second, I examine how Roberts drew upon a slightly different discourse of expertise when she presented the RDAs at a number of conferences apart from the NRC. The dominant genres in the field of nutrition and the NRC prevented citizens from actively engaging in and producing knowledge about nutrition, even within a system that purported to be "democratic."

DEVELOPING THE RDAS

When research on the RDAs began, committee members realized that they lacked rigorous scientific data for many nutritional values. In some cases, experts disagreed on how to measure a nutritional value scientifically, as was the case for vitamin A (Roberts "Usefulness" 109). In other cases, experts disputed the correct value for a given nutrition requirement. For instance, scientific estimates for ascorbic acid ranged from 30 to 120 mg, while estimates for thiamine ranged from 0.75 to 3 or 4 mg (Roberts "Scientific Basis" 59). Many vitamins had only recently been discovered, so scientists lacked data for these requirements. For example, Wilder noted that "Nicotinic acid was not even known as a vitamin until 1937, while the role of riboflavin was known only vaguely in 1935" (qtd. in Laurence 14). Thus, in some cases, Roberts could not rely on data published in scientific articles to construct her set of standards, but would have to rely on more informal or unpublished reports to supplement the missing pieces. Further, Roberts would need the members of the Committee on Dietary Allowances to agree on nutritional values despite the conflicting evidence for many of the nutrients. In order to ensure her committee's support, and ultimately that of the wider group of experts who would use the guidelines, Roberts enacted what she called the "democratic approach."

First, the committee solicited input from every scientist who had conducted research related to nutritional requirements. Each member of the committee reviewed published information on a set of nutrients, and Roberts enlisted graduate students, faculty members, and participants in her own classes at the University of Chicago to help with the search (Doyle and Wilson 73). The committee also sent requests to over fifty nutrition authorities for their input on any nutrient they had researched (Roberts "Scientific Basis" 60). After gathering all of this information, the committee devised a tentative set of allowances. Since scientific data was scanty for many values, the committee often had to estimate appropriate quantities of vitamins or nutrients. For instance, at the time, no experiments had been conducted to determine human requirements for niacin. Roberts noted later that the value for niacin in this first set of allowances was derived from a study of dogs, which was then checked against the quantity that was known to prevent pellagra in humans ("Scientific Basis" 64).

Second, this initial draft of allowances was submitted to all those who had contributed data, then to the NRC's Committee on Dietary Allowances, and finally to the broader Food and Nutrition Board. At each stage, Roberts modified the guidelines based on their suggestions. After another round of revisions, the committee drew up a final set of allowances, named them the Recommended Dietary Allowances (RDAs), and presented them to the committee in May of 1941. In all, forty-three scientists contributed data used in developing the RDAs, twenty-six of whom were women (Harper 3701). By drawing upon these individuals to contribute their findings, Roberts increased the credibility of the RDAs themselves.

Records of the NRC's Food and Nutrition Board indicate how Roberts negotiated her position and asserted ethos as chair of the Committee on Dietary Allowances. In these archival documents, Roberts is always referred to as the "Chairman" of the committee—a reminder for the contemporary reader of the ways in which scientific institutions were set up, rhetorically at least, as male institutions. Despite her reputation as an authority in the field of nutrition, Roberts was obviously not male. In the absence of Roberts's own opinions or reflections, one can only guess at her colleagues' responses to her. One might assume that any woman in a position of authority would be a rarity in the 1940s, and that possibly some men would resent answering to a female chairman. Perhaps for this reason, Roberts did not enact a hierarchical approach, in which she privileged her own judgment over that of her committee members or other experts. Instead, the democratic style enabled her to distribute authority across the entire committee, and indeed, the field of nutrition. In this way, the democratic approach parallels what Karlyn Kohrs Campbell has called the "feminine style," insofar as it involved a personal tone, audience participation, authority based on experience rather than status, recognition of audience members as peers, and identification as a means of persuasion (13).[2] This approach enabled Roberts to construct a persona that would position her as fair, impartial, and nonthreatening, yet authoritative.

According to the minutes from a Committee on Food and Nutrition meeting held November 25th and 26th, 1940, Roberts first mentioned the idea of using the democratic approach to developing a nutrition conference, based on her experience with a previous conference on child welfare. On the second day of the meeting, the committee members discussed their plans to introduce the new dietary standards at a conference on nutrition, which would be held the following May. Roberts stated: "I think it is highly important that all the groups who have an interest in it, if, instead of handing them something, they have a chance to criticize it and be in on the planning and if we put out something to send to them for criticism and further suggestions with the idea that it will be modified" (National Research Council, *Minutes: Meeting of the Committee on Food and Nutrition* 22).[3] On the same day, Roberts presented an initial report of the Committee on Dietary Requirements—presumably, the standards she had drawn up with Stiebeling and Mitchell the night before (according to her 1958 speech). The minutes show that several changes were made based on an extensive discussion among the committee members. However, the official version of the RDAs was not released before another six months of research and consultation. Roberts's rhetorical task throughout this process was to ensure that her fellow committee members saw her as a fair and impartial chair, and that they respected her judgment and methods.

To accomplish this task, Roberts makes explicit mention of the democratic approach in her regular reports to the Food and Nutrition Board. For instance, she circulated a draft of the allowances in a letter report to Wilder dated December 31, 1940, which was later distributed to members of the Committee on Food and Nutrition as a circular letter. The report begins with a disclaimer:

> I sent out about 25 letters to specialists who had a right to an opinion on requirements, and the reports are just beginning to come in. I had hoped to have time to go over them more carefully and try to set up standards which would be in harmony with the various points of view, so far as this is possible. There are, of course, some very wide differences of opinion and these are going to be rather hard to reconcile. What I had hoped to do was to form some allowances on the basis of original material sent in, returning them to the contributors for criticism and then re-vamp the allowances. I still think this ought to be done, and it can be if there is not too much rush about getting them in. (*Circular Letter No.2* 38)[4]

In this preamble, Roberts diminishes her own role as compiler of the allowances, emphasizing instead her dedication to finding compromises and consensus that would ensure widespread support. This opening helps to shape Roberts's ethos as a judicious, fair, and impartial compiler of data, one who wished only to represent the views of other experts to the best of her ability.

Roberts continues to develop this persona throughout the report. She provides evidence for each of the nutritional values included, showing which data was used and why, and occasionally provides her own opinion. For example, she writes regarding the values for vitamin A: "They are not as high as many people tend to set them, and the Committee will probably want to raise them. Personally, I think it a mistake to put the vitamin A requirements too high because it seems to minimize the value of milk, eggs and other dairy products. As a matter of fact, I believe most workers would agree that if a person has a pint to a pint and a half of milk, some butter and an egg, he would more than have his vitamin A requirements met" (40). While Roberts does introduce her own opinion in this passage, she frames it as a perspective with which "most workers would agree" (40). In this way, Roberts positions the value she selected based on the evidence as a prudent and pragmatic one that would gain the assent of her colleagues. The report also features tables showing the exact numerical values posited by different researchers for each value, along with Roberts's recommendation. By providing this kind of detailed evidence, Roberts is in effect "showing her work," thereby giving members of the Food and Nutrition Board everything they needed to determine that Roberts's own judgments were suitable.

Roberts also enacted this persona in face-to-face meetings, which is evident from a transcript of a meeting held in 1944. The meeting in question was held to determine what, if any, changes were warranted based on research that emerged in the three years after the RDAs were first released. At this time, Roberts determined that enough new evidence existed to reconsider the original allowances, particularly the guidelines for thiamine and riboflavin. As chair of the committee, Roberts opened the session with an overview of how the RDAs had first been developed, insisting on the fact that they were developed collaboratively, and that they would be revised whenever sufficient data suggested this was appropriate. The chair of a meeting often begins by outlining an agenda, and Roberts did just that, suggesting that the committee discuss general principles first, before moving onto specific nutrients. However, Roberts also gave her audience members an opportunity to suggest other courses of action: "If you think any other order of business would be preferable, I should be happy to have you say so; that is, if you think we should go right to the discussion of the pieces of evidence, or if you think the basic topics that I have outlined should have a different order, I should be glad to have you say so" (National Research Council *Proceedings: Committee on Dietary Allowances* 8).[5] In this way, Roberts enacted some qualities associated with the democratic approach, treating her committee members as peers and encouraging their participation in shaping the course of the meeting. Throughout the meeting, Roberts solicited opinions from the committee members, occasionally offering her own viewpoint but more often acting as a facilitator for the discussion. Roberts ends the meeting with the following caveat: "May I say once more than I think you will all appreciate, in view of this discussion, that the allowances that will be finally set up will not be such as every single person would prescribe to, but we will do the best we can to formulate one set which will represent the consensus of the group and at the same time be practical" (226). Thus, throughout the meeting, Roberts worked to ensure that everyone present could provide input, and that everyone, therefore, would be likely to accept the final allowances as the best compromise possible based on the range of opinions presented. This did not mean that Roberts completely effaced her role as chair, or that she declined to present her own opinions. The democratic approach did mean, however, that she was able to claim authority in a way that would seem acceptable to a mixed group of male and female experts.

PROMOTING THE RDAS

The negotiations over the RDAs in internal NRC documents show that the values adopted were very much in dispute and subject to negotiation. Further, persuading experts to accept the RDAs would be difficult because, while a few nutritional standards already existed in 1940, there was no single authoritative standard to guide their efforts. In a later speech, Roberts summarizes the state of affairs: "It is true that there were many dietary standards already in use—too many of them, in fact. But these varied greatly in the values set for various factors and in the degree of acceptance accorded to them by groups needing such standards" ("Usefulness" 105). For example, Lucy Gillett had worked out calorie standards for children, Henry C. Sherman had developed standards for adults for four nutrients (protein, calcium, phosphorus, and iron), and Stiebeling had proposed a set of standards for the United States Food and Drug Administration (FDA) (Doyle and Wilson 72).[6] Other sets of standards had been proposed in the 1930s by the League of Nations Health Organization and the Canadian Council on Nutrition (Harper 3700). Roberts's committee would have to persuade nutrition experts to adopt the NRC's standards, rather than the ones that were already in place.

When Roberts presented the RDAs to scientific audiences, she positioned the NRC as the authoritative opinion of a group of experts. Here, the democratic approach serves the rhetorical function of shoring up the authority of the NRC, rather than opening up authority to others interested in nutrition. In this section, I examine how Roberts established the RDAs as "expert knowledge" in speeches she made in 1943 and 1944. Roberts presented the first of these speeches, "Scientific Basis for the Recommended Dietary Allowances," before the Annual Meeting of the Medical Society of the State of New York in 1943.[7] Roberts delivered the second speech, "Usefulness and Validity of the Recommended Dietary Allowances," at the Norman Watt Harris Memorial Foundation's Twentieth Institute in Chicago, September 1944. The goal of the conference was to assemble leaders in nutrition, international relations, food supplies, and population to discuss matters of food policy (Schultz ix).[8]

When she gave presentations on the RDAs, Roberts tended to employ three fundamental strategies. In fact, Roberts's speeches about the RDAs were often quite similar in structure and content, although of course she adjusted some minor details for each occasion. For this reason, it seems useful to consider her RDA presentations together in order to tease out some of the common rhetorical strategies she used.

First, Roberts established the collective authority of the NRC. As Hilgartner has argued in his analysis of more recent NRC reports, when preparing advice, advisory bodies do not only review evidence and present recommendations; they also present the advisory group itself as a character—particularly, as a scientific organization of great prestige (44). When she spoke or wrote about the RDAs, Roberts represented not only her own interests, but also those of the NRC. Accordingly, she took pains to establish the credibility of the NRC as a whole, a group she described as "a body whose reputation for scientific caution and soundness is unparalleled" ("Usefulness" 106). She also praised the NRC's foresight in recognizing that nutritional standards would be needed in the first place: "The need for such allowances was recognized by the board shortly after it was organized in the fall of 1940. The board realized that one of its functions would be to advise on nutrition activities of many types during the war period and after. Indeed, that was the purpose for which it was created. It knew that to do this task wisely it should have some sound nutritional goals at which to aim" ("Usefulness" 105). In her speeches, Roberts presented the NRC as authoritative, scientifically rigorous, and prescient.

Second, Roberts described the democratic approach itself, emphasizing that the RDAs presented the collective agreement of a wide group of experts. In this way, the democratic approach actually served to shore up the authority of an elite group of scientists. In her "Scientific Basis" speech, Roberts wrote: "It is clear then that the allowances had to be set up partly on the basis of *the judgment of nutrition workers whose experience gave them some basis* for deciding what values to interpolate for those that were lacking" (65, emphasis added). Similarly, in her "Usefulness and Validity" speech she attributes the success of the RDAs to the fact that they were put out by the NRC, and to "the fact that they did not represent the opinion of one individual or small group, or even of merely the members of the board, but were rather the considered judgment of *a large group of nutrition authorities*" (106 emphasis added). In other words, Roberts made it clear that the RDAs did not reflect her own judgments, but those of a large panel of experts. In this way, Roberts bolstered the credibility of the RDAs and the NRC, deflecting attention away from her own role in developing them.

Third, Roberts describes the scientific basis for each nutrient included and indicates where future research might fill in gaps in the knowledge base. By framing the RDAs with relation to future research opportunities, Roberts encourages the participation of her scientific audience in shaping the RDAs. For instance, in her "Usefulness and Validity" speech, Roberts takes up each value represented in the RDAs in turn, explaining the scientific evidence for each. Roberts claims that the caloric allowances would probably not change much, since they were based on plenty of sound research, but that more research was needed in relation to calcium requirements, especially for adolescents ("Usefulness" 108). In all, of the eight vitamins Roberts reviewed, she expected definite revisions for two: the A and B vitamins (131). In this speech, Roberts also notes that because the RDAs had already helped to stimulate research: "The review of the evidence which the board had carried out revealed great gaps in our knowledge of what the requirements are for the different factors, and workers in many laboratories have been challenged to do research to fill in the gaps" (107). In this way, Roberts shows that the RDAs illuminated gaps in current knowledge, creating opportunities for others. By extension, this call for participation places the RDAs solidly in the center of nutritional research and also encourages her audience to identify with the interests of the NRC.

Overall, Roberts's rhetorical use of the democratic approach among scientific audiences serves primarily to support the objectivity of the RDAs and the expertise and authority of the NRC. The term *democratic* here did refer to a participatory process, but one in which only a select group of scientists were included.

PUBLICIZING THE RDAS

The RDAs provided a common starting point for wartime nutrition advice from experts to broader public audiences—including applied nutrition workers (home economists, dietitians, extension workers, and the like) and citizens who might use them in everyday food planning. In her "Usefulness and Validity" speech, Roberts notes that by 1945, the RDAs were already being used for a wide range of purposes: to guide nutrition enterprises in the United States, Canada, and England, to plot menus for the Quartermaster Corps of the U.S. Armed Forces, to plan relief feeding in occupied countries, to advise the population on securing an adequate diet, to set up industrial feeding programs, to describe nutritional requirements in standard textbooks in biochemistry, nutrition, and pediatrics, and to prepare state and local nutrition programs.

Despite the democratic procedure used to develop them, the RDAs were positioned within a scientific culture that produced a top-down flow of information, so that those who were ultimately meant to benefit from them did not have a say in how they were developed. The RDAs were based on scientific data gathered from research scientists, and were also revised periodically based on new scientific information. In turn, the RDAs were packaged and redistributed for a number of additional audiences, but these groups had little, if any, direct input into new revisions of the RDAs. The research scientists—those who were well positioned like Roberts, Mitchell, and Stiebeling—stood to gain the most, professionally, from this system. Indeed, after the RDAs were published, and throughout the war, Roberts, Mitchell, and Stiebeling were featured in newspaper articles about nutrition. In this way, the RDAs helped to maintain and increase the status of these scientists. Stiebeling, for instance, appeared in several *New York Times* articles during the war. In one case, she is quoted as warning of impending wartime shortages in fats, vitamin A, and carbohydrates ("Warns of Shortage of Fats in Our Diet"); in another, she is profiled (with a photograph) after her appointment as chief of the Bureau of Human Health and Home Economics in 1944 ("Dr. Stiebeling Gets U.S. Nutrition Post"). The latter article makes Stiebeling's status as an expert explicit ("Dr. Stiebeling has for fourteen years been a leading expert of the bureau") and mentions her wartime work on meal planning (34).

The first version of the RDAs was released in May 1941. This table appeared not only in the NRC's own booklet, but was also reproduced in textbooks and journals. Scientists used them to identify gaps in existing research, applied scientists used them to develop educational materials, and dietitians used them (often in a mediated form) to plan for food purchases and preparation.

The RDAs also provided rhetorical fodder for experts seeking to change nutrition policies. For example, at the National Nutrition Conference for Defense, held in May 1941 in Washington D.C., Stiebeling presented the findings of a study she conducted on the dietary situation in the United States. Stiebeling compared the results of two large-scaled studies of diets in the United States with the RDAs. She concluded that, based on surveys of the amount of money families spent on nutrition, "35 percent of those in cities and in villages had poor diets as compared with 25 percent on farms" (*Proceedings of the National Nutrition Conference for Defense* 85). Stiebeling's studies were taken up elsewhere to justify proposals to improve nutrition in the nation. In his address to the American Dietetic Association in 1941, Wilder noted that Stiebeling's work demonstrated that well over 45 million Americans could be considered malnourished if their food consumption was compared with the RDAs (2). Stiebeling's figures were often cited as proof that the United States had a long way to go before it would reach the goal of adequate nutrition for all. They were quoted in Icie G. Macy and Harold H. Williams's 1945 textbook, *Hidden Hunger* (210), in Nelda Ross's 1941 presidential address to the American Dietetic Association (ADA) ("Editorial: The American Dietetic Association and the Defense Program" 790), and in Wilder's presentation at the opening session of the annual ADA meeting in October, 1941 (2). The RDAs provided an opening for experts to conduct further research and expert discourse.

Roberts herself appealed to the RDAs in order to push for policy changes. In a 1944 article entitled "Improvement of the Nutritional Status of American People," Roberts argued for measures that would help to distribute food more equitably in order to prevent deficiencies such as pellagra, beriberi, scurvy, and rickets ("Improvement" 401). These measures included mandating iodized salt, enriched flour, and fortified cornmeal; rationing sugar; and funding a national school lunch program (401–3). Roberts noted, further, that the war provided a most effective rhetorical moment in which to lobby for such changes. For instance, in terms of flour enrichment, she wrote: "After the war the big millers and bakers will probably continue to enrich without compulsion, but many of the smaller ones will not. *Therefore it is essential that legislation be passed by the states, and the psychological time to secure that legislation is now.* Some 40 states have sessions of their legislatures in January 1945, and now is the time for home economists, state nutrition committees, and all interested groups to begin to work for such legislation in their states" (402, emphasis in original). Many of these measures were successful—iodized salt, enriched flour and cornmeal, and the school lunch program persist even today. The RDAs provided a common, well-accepted rhetorical basis for arguments about nutrition policy.

The RDAs generated a large set of documents for nutrition workers, including textbooks, pamphlets and brochures, government bulletins, and tables and charts meant for students and professionals in nutrition. For example, the United States Department of Agriculture issued a bulletin entitled "Planning Diets by the New Yardstick of Good Nutrition" in 1941, meant for "nutrition specialists, teachers, social workers, public health officers, nutrition committees, land-use committees, or administrators concerned with problems of food purchases or food production for families, a community, or the nation" (Bureau of Human Nutrition and

Home Economics *n.p.*). The RDAs were also reproduced in educational materials targeting the American homemaker. These include books, such as Alice Frances Pattee's *Vitamins and Minerals for Everyone* and Morris Fishbein's *The National Nutrition*; brochures, such as the one produced by the Philadelphia Child Health Society called *Family Nutrition*; and advertisements.[9] These documents often stress visual data and categories of foods, rather than the quantitative data found in the tables distributed to nutrition experts. They also stressed advance planning, providing worksheets to help homemakers plan appropriate quantities of food to purchase and menus for meals that satisfied the RDAs.

These documents emphasized the authority of the NRC and the validity of the RDAs. As is common in scientific discourse for popular audiences, these texts did not involve the public in determining nutritional values or their format, but instead focused the reader's attention on applying those values to their everyday lives. This advice literature commonly positions the RDAs as authoritative, expert opinion, echoing the rhetoric used by Roberts, Wilson, and others. For example, Fishbein wrote, "This chart is now being called the yardstick for nutrition. The recommendations are based on a study of all of the available scientific literature and on opinions from a number of authorities on nutrition." (18). By replicating Roberts's own emphasis on the RDAs as the expert judgment of nutrition authorities, Fishbein limited the scope of the audience's concerns to issues of application—how best to use the "yardstick" in planning meals.

News media echoed the approach found in the advice literature, adopting the epideictic, or celebratory, mode Jeanne Fahnestock has identified as key to scientific news journalism. The RDAs made the front page of the *New York Times* on May 26, 1941, the day after they were announced at the National Nutrition Conference for Defense. In the article, those attending the National Nutrition Conference praised the committee's success. Paul V. McNutt, the Federal Security Administrator, stated that the new standards would set nutritional standards toward which government, industry, labor, science, and education groups (along with individual citizens) must work "if America is to realize her full strength for defense" (qtd. in Laurence 14). Thomas Parran, the United States Surgeon General, stated that "Now, for the first time, the United States has definite nutrition recommendations from an authoritative national committee which has pooled all the available knowledge on foods and drawn a blueprint of the amounts and kinds of dietary essentials for good health" (qtd. in Laurence 14, emphasis added). News media helped to sell the RDAs as an authoritative food guide, once again limiting the extent of public participation to application only.

Margaret Mead: Changing Food Habits

The RDAs became an integral part of the national nutrition program during World War II. Dietitians working in schools, hospitals, and industrial cafeterias had to adjust the guidelines for individuals, based on factors such as age, weight, activity level, and the specific foods available in each community. To popularize the RDAs for the general audience, nutrition experts conducted demonstrations and wrote newspaper columns, magazine articles, and informational bulletins. The RDAs helped to shape and coordinate this effort not only in the United States, but internationally, as nutritionists planned for eventual food relief efforts after the war. Throughout these efforts, nutrition workers drew upon a discourse of expertise to assert their place within the "second line of defense."

As the war progressed, this expertise became increasingly necessary as food shortages made day-to-day planning of nutritious meals increasingly difficult. Rationing was introduced in 1943, and meats, fish, fats, oils, cheese, eggs, milk, sugar, coffee, tea, fresh fruits and vegetables, and other commodities were often in short supply ("Rationing" 290). Dietitians and nutritionists had to devise meal plans that would conserve these precious items or introduce unfamiliar substitutes, like soybeans, without sacrificing nutritional quality. By 1943, nutrition experts had to revise their initial assumption that food shortages would not be a problem in the war. One editorial in the *Journal of the American Dietetic Association* (*JADA*) stated: "we come to 1943 and face with something of a shock the fact that food shortages promise to be fully as acute, if not more so, than in 1917–1918 [. . .] In fact, for the first time in our history we are faced with the need for over-all rationing" ("Editorial" 110). Although dietitians and nutritionists recommended substitutes for rationed ingredients such as sugar, milk, and meat, the editorial noted, "once a food is suggested as a substitute for another [. . .] it is but a question of time before the substitute, in turn, becomes scarce" (110). Nutrition experts were increasingly called upon to help Americans adjust to scarcity and to unusual new foods.

Even though the war presented temporary problems—rationing, food and labor shortages, and so on—dietitians and nutritionists viewed the war as an opportunity to stimulate long-term improvements in American dietary habits. In one article, Mitchell claims that the "present emergency" had "jogged us out of a rut and made us think what we who know fundamental nutrition facts can do toward applying them. It has made the laymen who were formerly disinterested or indifferent more willing to listen to suggestions and make application. Now we need to take advantage of these circumstances to give people the information they are seeking in a form they can understand and use" (538). Given the growing recognition of nutritional issues, professionals in the field sought to capitalize on the opportunity to improve the nutrition of the American population and to improve their professional reputation.

However, changing food habits was no small feat. Jennie Wilmot claimed in a 1943 article that food habits were deeply engrained in history and custom: "Even you and I who supposedly are aware of this and open-minded with respect to food, have our likes and dislikes, our more or less ingrained habits and perhaps prejudices. Food customs change slowly" (505). Throughout the war, dietitians and nutritionists engaged in a continual struggle to use their expertise to enact changes in food habits.

The counterpart to the Nutrition and Research Council's Food and Nutrition Committee, which addressed biochemical and physiological aspects of nutrition, was the NRC's Food Habits Committee, which addressed cultural and psychological aspects of food. Although the Food Habits Committee provided advice mainly to home economists and dietitians, its membership was drawn mostly from scholars in cultural anthropology, psychology, sociology, and nutrition.[10]

In 1942, Margaret Mead took a leave of absence from the American Museum of Natural History, where she was Associate Curator of Anthropology, to serve as executive secretary for the Food Habits Committee. Mead had earned a national reputation in anthropology, due to her groundbreaking studies of indigenous cultures, including *Coming of Age in Samoa* (1928) and *Sex and Temperament in Three Primitive Societies* (1935). Mead's work, in particular, had contributed to a move toward culturally relativistic, as opposed to evolutionary, approaches to anthropology (Newman 236). However, Mead had also gained a wider public reputation, in part by writing articles in popular magazines such as *Vogue, Redbook, Good Housekeeping,* and the *New York Times Magazine* (Newman 233). Thus, she could take advantage of her reputation as a scientist beyond her own discipline to direct the wartime study of food habits. This section first explores the rhetorical strategies Mead used to claim credibility for an anthropological approach to nutrition work, and then examines how this strategy duplicated structures of expertise without granting additional agency to the cultural groups being studied.

JUSTIFYING THE ANTHROPOLOGICAL APPROACH

The first task for Mead and her committee was to justify an anthropological study of food habits, a problem normally addressed by researchers in nutrition. Since the anthropologists and social scientists on Mead's committee were not formally trained in nutrition, home economics, or dietetics, the committee needed to build credibility among audiences made up primarily from those disciplines. Mead employed two main strategies to justify the contribution anthropologists could make to the issue of food and nutrition in articles and reports addressed to those audiences. These documents included "The Problem of Changing Food Habits" (a committee report published in 1943), "Dietary Patterns and Food Habits" (published in *JADA* in 1943), and the *Manual for the Study of Food Habits* (published by the committee in 1945). Mead argued that an anthropological perspective could balance the short-term perspective of dietitians and nutritionists with a deeper understanding of food habits in their cultural context. Given their grounding in studies of culture, Mead argued that anthropologists could help to identify which food habits were relatively permanent and which were more temporary. Like Roberts, Mead draws on a discourse of expertise to argue for the usefulness of anthropological research for the "second line of defense."

When she addressed audiences composed of nutritionists and dietitians, Mead argued that an anthropological perspective could help to correct the short-term focus of those disciplines by providing more in-depth, long-term analyses of food habits. Mead implicitly suggested that disciplines such as nutrition, dietetics, and home economics were shortsighted in their research, and that anthropological perspectives offered a deeper, richer perspective. For instance, in a presentation to the American Dietetic Association on October 19, 1942, "Dietary Patterns and Food Habits," Mead described two main ways in which the Food Habits Committee could address food issues during the war. First, Mead noted, anthropologists, sociologists, and social psychologists could "place any set of activities within our society against the backdrop of a systematic knowledge of many other cultures" ("Dietary Patterns and Food Habits" 1). Further, anthropologists could "give a knowledge of the cultural dynamics which underlie our present acceptance or rejection of certain dietary practices—as a people. By identifying the peculiar culturally standardized attitudes which have been built into us, as members of the American cultural group, it will be possible to develop a dietary pattern of greater durability than is the case when the dietitian has to operate on an intuitive, trial and error basis" (1). In other words, Mead claimed that anthropologists could demonstrate how knowledge of cultural traditions and patterns could inform solutions to wartime food problems.

In order to change food habits, she argued, dietitians needed to understand the cumulative process by which children learn to eat, how they build up attitudes toward food, and how their personalities are involved with eating (3). In the United States, Mead suggested, children are taught to associate food with morality: they are rewarded with the "wrong" foods (sweets or treats) for eating the "right" foods; if they do not eat the "right" foods, children are threatened with missing dessert. In this way, Mead claimed, "A permanent conflict situation is established which will pursue that child through his life—each nutritionally desirable choice is made with a sigh or rejected with a sense of guilt; each choice made in terms of sheer pleasure is either accepted with guilt or rejected with a sense of puritanical self-righteousness" (3–4). Mead argued that these patterns of acceptance and rejection must be addressed when making recommendations to the American population.

In a 1943 report of the Food Habits Committee, Mead offered another justification for anthropological approaches. Here, she claimed that focusing on short-term needs alone could have negative long-term effects. Mead wrote that the committee's task was to estimate "clearly the relationship between any particular change which may be introduced in the dietary patterns, or the culturally standardized methods of inculcating food habits, and the impact of such a change on other parts of our culture or the culture of other peoples of the world with whom we come into contact through lend-lease, relief and rehabilitation procedures" ("The Problem of Changing Food Habits" 20). For example, relief feeding efforts might distribute white flour to populations that traditionally ate whole grains. Given the nutritional deficiencies of white flour, developing a taste for it would inculcate "a habit with disastrous repercussions for their future health" (20–21).

According to Mead, the anthropological approach provides a deeper perspective because it situates food habits within the broader dynamic of a cultural tradition. Using this approach would guard against taking measures that might be desirable in the short term, but disastrous in the long term: "Only by putting each recommended innovation and the methods suggested for bringing it about against the total cultural picture, is it possible to guard against initiating changes which, while nutritionally desirable in the narrow sense, may be socially undesirable in a wider sense" (21). In this way, Mead established the validity of her anthropological method to the issue of food habits, normally under the purview of dietitians, nutritionists, and home economists.

In addition to its greater emphasis on cultural patterns, Mead contends that anthropological approaches could help researchers to distinguish elements within a culture that are relatively permanent, or stable, from those that are more susceptible to change. Even as she asserts the significance of this approach, Mead makes sure not to alienate her audience of nutritionists and dietitians. She suggests that the extraordinary circumstances of war provided a chance for all nutrition workers (including anthropologists) to shift food habits. Perhaps because food customs and habits are so deeply embedded in past behavior, Mead suggests, the unique constraints of wartime offer "a tremendous opportunity for the institutional dietitian to take the leadership in establishing a public opinion which will demand well-balanced meals prepared in a way which conserves the nutritional values of food" ("Dietary Patterns" 4). Such a change in public attitude could help to establish a more permanent pattern, one in which meal planners would be targeted with nutritional information to help them prepare more healthful meals.

Drawing on Mead's arguments, nutritionists argued that they could use their expertise to encourage Americans to exchange foods with poor nutritional value, such as sweets and sodas, for vitamin-rich fruits, meats, and vegetables. Although such poor food habits may have been serviceable earlier, they were no longer appropriate for a situation in which every citizen needed to be healthy and energetic to help support the war effort, and in which rationing limited sugar supplies. Wilmot recommended that dietitians and others involved in nutrition education might take advantage of the war to "teach people gradually to adjust to one sweet dessert a day instead of the accustomed two. Perhaps the use of more fresh fruit as a conclusion for meals, in time, will become more general. It may be that we'll learn to prefer the sharp perfection of good black coffee, and mellow, well-brewed tea, and the wholesome goodness of cereal and milk, without the urge to drown the distinct flavor of each with cloying sugar" (506). The specific exigences of wartime, such as rationing, created an opportunity for people to give up long-standing habits and to develop new ones. As Mead suggests, the wartime crisis meant that "We can accomplish a major cultural shift: substitute a moral premium on demanding and serving balanced nutritional meals, for our traditional, non-scientific moral premium on eating foods branded as unpalatable" ("Dietary Patterns" 5).

For Mead, anthropologists' expertise extended to rhetorical strategies as well. Because societies change quickly and continuously—along with demographics, food habits, and scientific knowledge of nutrition—the effort to educate the public about food and nutrition must also adjust to changed conditions (NRC, *Manual for the Study of Food Habits* 17). The rhetoric used in this effort should change as well, as Mead notes, since "Appeals to motivations appropriate at one time and place will prove empty and dead at another time and place" (16). However, this focus on change must be balanced with an understanding of which cultural elements remain static. Mead argued that anthropological research could provide this type of focus.

Overall, Mead suggests that anthropologists could help to shift the research focus for nutrition. The fundamental question for this research program would shift focus away from the question, "How can we change food habits?" that had previously informed such studies. This question, according to Mead, presupposed a level of fixity that was not congruent with how food habits function in society: "'How can we change food habits?' assumes that a given population has a fixed set of food habits, some of which are incorrect by our standards, and so we wish to change them to an equally fixed set of habits which will conform to the science of nutrition" (24). Instead, research in food habits should begin by developing basic knowledge about the nature of nutritional behavior, and then shift to the following question: "How can we develop food habits which have the requisite stability and flexibility appropriate for given individuals in a given society at a given time" (25).

Thus, Mead argues that anthropological methods could help nutritionists to balance a concern for change in the present with a concern for the more stable features of a given culture. Anthropological research could supplement the short-term, present-oriented focus of nutrition research with knowledge of long-term cultural habits and patterns. Mead's outline of the advantages of her approach, coupled with the rhetorical force of "total war," helped to ensure the success of the Committee on Food Habits, despite the unorthodox training of many of its members. Relying on a discourse of expertise ensured the place of anthropologists among other experts on nutrition.

CULTURAL EXPERTISE IN ACTION

The suggestions made by the Committee on Food Habits were carried out in a number of ways. The committee supported research into the food habits of a number of American population groups, much of it undertaken by Natalie Joffe, who worked with members of the group in question to develop a more nuanced study. For example, Joffe and Tomannie Thompson Walker prepared a report entitled *Some Food Patterns of Negroes in the United States of America and Their Relationship to Wartime Problems of Food and Nutrition.* The report follows the suggestions put forth by the Committee on Food Habits, situating specific practices in relation to culture, socioeconomic status, family organization, and geographic location, describing in detail the differences and similarities between African-Americans living in the rural south, urban south, and urban north. The program of research Mead advocated and Joffe and Walker adopted allowed certain representatives of ethnic groups in America to speak as experts for their communities. While this expertise may have resulted in more culturally relevant nutrition advice, it did not necessary empower members of those cultures to take an active role in shaping nutrition policy. Instead, it kept in place a system of expertise that privileged the perspectives of nutrition workers over those who were expected to listen to them.

Studying food habits led to some important insights that might help nutritionists to translate information into culturally relevant terms. Joffe and Walker's report goes beyond stereotypical assumptions about African-Americans as a group, taking into account changing conditions in each region—education, socioeconomic status, and so on—and their effects on food habits. For example, the report differentiates between African-American women in urban and rural settings; north and south; and upper, middle, and lower income groups. Joffe and Walker also noted that food habits were connected to deeply engrained cultural values: "Until recently it has been axiomatic to consider the slaves imported from Africa as cultural paupers, stripped of their traditional heritage, and that whatever their behavior, it derived from conditions operating in the New World. However, within the last few years this position has been reversed and the hunt for such survivals has yielded rich quarry. Far from being the clean slate, the immigrant 'African' brought a rich heritage of culture. These heirlooms were by and large intangibles and had been retranslated into many phases of activity on the new soil" (27). The writers mention a few of these cultural inheritances, including culinary terms such as *gumbo* and *okra,* and the tendency to use food as a gift when a guest in another's home (27). Joffe and Walker also warn against the kinds of demeaning and stereotypical images that might have been used in publications for African-American audiences. For instance, they tell experts to avoid pictures of "the 'Aunt Jemima' type, which means no bandanas on the heads of women who are neither young nor slim. Women should be neat, prettily but not too brightly dressed, and in decent surroundings" (24).

However, Joffe and Walker's report also points to some of the more troubling aspects of outreach programs in nutrition without fully addressing their implications. For example, they note that in many cases, individuals learned nutrition information without putting it into practice: "Interviews with relief clients often show that a woman can rattle off a completely balanced food list and the reasons why she is buying these foods, when it is known that she is not practicing what she has been taught. A response of this type can be expected because the woman is dealing with organized 'authority.' Many submissive persons exhibit this type of passive resistance" (23). But the researchers do not address some of the possible reasons behind this kind of resistance. It seems likely that many individuals (not just African-Americans) might resist nutrition advice that seemed to be imposed on them by outsiders. In African-American communities, especially, this resistance may have stemmed from a long history of imposition from experts seeking to "help." Joffe and Walker do not go so far as to suggest that nutrition educators involve members of the community in developing outreach programs and materials; instead, they simply duplicate hierarchies of expertise. They suggest, for instance, that nutrition educators create alliances with other experts in African-American communities: "The personality factors of ministers, teachers, physicians, and other local persons of repute may spell success or failure of any project. In certain places it may be necessary to secure the approval of local Whites, or for them to be the titular heads of programs, while at other levels Negroes may be more efficacious. In dealing with men, a man might be better, while for women, men or women workers are more productive depending upon the situation" (23). Thus, while Joffe and Walker's study provides important insights into cultural food patterns, it does not dislodge or challenge existing, hierarchical models of nutrition outreach.

The Discourse of Expertise

Overall, Roberts's democratic approach and Mead's anthropological approach to nutrition outreach limited knowledge-making to scientific elites. Citizens had little say in how guidelines were developed, and for what purposes. For this reason, valuable insights from American citizens, especially homemakers, were overlooked. For instance, in a 1943 study, Earl Lomon Koos found that low-income, urban women resented nutritional guidelines that required them to plan meals on a weekly basis. Instead, women preferred to shop for foods every day, both as a way to escape the confines of a small apartment, and as a way to exercise choice. Further, Koos found that people were more receptive to programs targeting individual foods (i.e., drink more milk) rather than wholesale patterns (Lomon Koos 80). Because certain foods indicated status for some cultural groups (such as white bread or red meat), many women resisted being told to purchase whole wheat bread or cheaper cuts of meat (79). Yet, the advice literature that circulated during the war rarely, if ever, included these insights.

For the most part, pamphlets and brochures included tables that encouraged homemakers to plan weekly meal allowances based on the RDAs, which were described as a "yardstick" for good nutrition. Not surprisingly, the kinds of meals suggested reflected primarily Anglo-American food preferences. For instance, the first edition of the RDAs included a sample low-cost meal plan (shown in Table 4.1) that featured pot roast, potatoes, and plenty of milk. Clearly, this sample menu overlooks the fact that Americans from some cultural backgrounds might not eat pork (or any meat at all), that many others might prefer pasta or tortillas to bread and potatoes, and that high proportions of nonwhite individuals are lactose intolerant. Despite the research undertaken by the NRC's Committee on Food Habits that sought to identify patterns and customs for ethnic enclaves in America, this edition of the RDAs does not mention specifically that these guidelines should be translated into culturally appropriate meal plans. Instead, it suggests simply that they should be used by nutrition experts in accordance with the "appropriate quantities of foodstuffs available in their localities" (NRC, Committee on Food and Nutrition 2).

Further, nutritionists did not use their expertise to encourage men to participate in these activities. Even though the American housewife was encouraged to enter war industries, those whose husbands were also working on the home front were still expected to perform their traditional domestic duties, balancing meals based on the RDAs all the while dealing with rationing and food shortages. Thus, the RDAs shaped other genres of information along axes of status and gender. Far from a neutral set of scientific standards, they enacted and encoded a discourse of expertise.

Finally, some critics argued that the wartime focus on nutrition, promulgated in large part through the RDAs, actually limited the scope of women's involvement in the war, channeling their energy into tasks that were simply extensions of their prewar domestic duties. Susan B. Anthony II wrote in 1943: "Lurking behind the nutrition posters, the com-

Table 4.1. Sample menu from the Recommended Dietary Allowances, published by the National Academy of Sciences, 1941.

Breakfast	Lunch	Dinner
Tomato juice	Baked navy beans	Pot roast and gravy
Oatmeal with top milk	Cabbage salad	Baked potatoes & oleo
Toast with oleo	Bread with oleo	Carrots
Coffee for adults	Prunes	Bread with oleo
Milk for children	Milk	Tea or coffee for adults
		Milk for children

mittees and conferences, was the hoary notion that in the solemn business of winning a war women's chief contribution should come through *more* hours of cooking, *more* hours of shopping and *more* conversation about food, meat cuts, vegetables and vitamins. Naturally, no one would minimize the importance of food in war time as well as peace time; but, from the point of view of serious women's work the suggested confining of womanpower to the monotonies of nutrition seemed about as enlightened as limiting them to knitting" (43). While nutrition work provided professional opportunities for some women, it did so at the expense of other women, who were given even more detailed guidelines and higher standards for the same "busy work—amateur, low in prestige and low in returns—that women have been doing for generations" (45). Further, American women were targeted primarily by advertising (both via government propaganda and advertisements from commercial food companies) that situated them firmly in a domestic role; historians agree that the emphasis on domestic activities outweighed messages encouraging women to join the war effort outside the home.[11]

In a 1958 speech, Roberts ascribed the longstanding success of the RDAs to the method used to develop the RDAs: "I am sure that the main reason they have stood the test of time as well as they have is that they were developed by this democratic procedure" ("Beginnings" 904). Yet, one could just as easily attribute their success to the fact that Roberts successfully constructed the RDAs within a hierarchical scientific culture and a discourse of expertise, using the democratic approach to bolster the ethos of the NRC.

Far from a neutral set of scientific standards, the RDAs reified structures of technological elitism. The primary rhetorical function of the democratic approach was to win assent from those scientific groups and respect from the wider public. Despite Roberts's assurances that the RDAs were developed via a democratic approach, that procedure extended only to research scientists in nutrition, not the dieticians, home economists, and homemakers who would end up using them. Those who would use the RDAs in their work and home lives were not consulted; instead, they were positioned as consumers of information (and of food products). The RDA put in place a set of genres that ensured the efficient transmission of official information and advice. The Food Habits research conducted under Mead's program performed a similar function, making food advice more culturally relevant, perhaps, but without involving individuals directly in shaping that information. Overall, both efforts helped to position nutritionists as experts and individual Americans as passive consumers of advice and information.

Roberts's and Mead's work provides an important example of how alternative scientific methodologies can fulfill a variety of rhetorical purposes, some more progressive than others. On the one hand, the democratic approach offers a model for a collaborative, socially engaged form of science, and Roberts herself provides an early example of a scientist who forged alliances dedicated to the practical pursuit of the common good. Similarly, Mead's program of anthropological research seemed to offer a way to counteract the tendency toward "one-size-fits-all" nutrition advice with materials tailored to specific ethnic groups.

On the other hand, these practices are also problematic in some respects. Roberts also used the democratic approach to forge consensus among experts, shoring up her authority and that of the NRC. Since it was relatively uncommon for women to hold leadership positions in scientific institutions in the 1940s, Roberts may have faced extra scrutiny from her peers. The democratic appr1oach shifted the attention away from Roberts's own credibility, since she was presenting the opinions of a large group of experts. However, using democratic approaches mainly to increase one's own credibility or that of other elites can be problematic, especially if it occludes the interests and perspectives of the citizens who might contribute to scientific knowledge.

Similarly, Mead's efforts to produce culturally relevant nutrition advice failed to dislodge existing hierarchies that gave nutrition workers more agency than those they sought to help. Harvey Levenstein has argued that the Food Habits Committee was largely ineffective. First, the top-down programs Mead envisioned, wherein nutrition experts would directly advise individual communities, were incredibly time-consuming. There simply were not enough trained nutritionists available to reach a significant portion of the population (72). Further, the anthropological approach Mead espoused made experts hesitant about tampering with existing habits, given the centrality of food to cultural traditions (73). Both of these problems stem, in part, from the hierarchical model envisioned by the Food Habits Committee. The notion of imposing new food habits on a population is fundamentally hierarchical, one that fails to spread habits or ideas as efficiently as the viral model of a food fad or of the multipronged marketing campaigns employed by industry groups.

Overall, this chapter points to the inextricability of expertise from existing structures of power. For Roberts and Mead, it was only by embedding themselves within those structures that they could do their part to change them. Yet, even a scientist of Roberts's or Mead's stature could not fully avoid the typical, hierarchical structures of scientific institutions, especially when dealing with one of the most prestigious of these, the National Research Council.

Perhaps because their expertise within the broader university system was shaky, women in nutrition and home economics were likely to abide by traditional models of scientific research and publishing to gain status. While they did more outreach work than some other scientific disciplines, publishing materials for public audiences on applied topics, they did not completely disrupt the one-way flow of information from scientific expert to consumer, nor did they challenge the discourse of expertise. Instead, they relied upon it to gain professional standing. This chapter suggests that simply increasing the number of women in a discipline will not automatically bring about structural changes to encourage more democratic forms of knowledge-making that feminists might anticipate. Rather than organizing the discipline in ways that would provide equal input from experts and nonexperts, the disciplines of home economics and nutrition were beholden to extant genres and institutional cultures of science.

Conclusion: Regendering Scientific Cultures

The examples of female scientists in this book demonstrate that women successfully appropriated scientific discourses and genres during World War II, making important contributions to science and the success of the war itself. Women such as Alice Bryan, Florence Goodenough, Dorothy Thomas, Leona Marshall, and Lydia J. Roberts used scientific discourses to their advantage, forging careers as exceptional women, often in male-dominated disciplines. Yet other women were limited by genres and discourses of scientific writing that muted their critical insights and ethical concerns. Georgene Seward, Katharine Way, Tamie Tsuchiyama, and Rosalie Hankey struggled to assert critical and ethical insights in a language that would be appropriate within their scientific fields, but their potentially transformative insights into dominant practices were suppressed.

Following Kristen Campbell, I suggest here that one goal for future research in science studies should be to account for the rhetorical practices that might result in more reflexive, situated accounts of scientific knowledge (178) and to generate new "tropes, discourses, or figurations of science" (178). Inquiries into the rhetoric of science should prove central to such an effort, since rhetorical studies map relations of power and gender alongside discursive practices. I use the term *regendering* to describe this effort to open up scientific discourses and genres.[1] Regendering scientific rhetoric would involve, first, identifying how scientific genres and discourses have developed in accordance with male-dominated cultures and disciplines, and second, identifying ways to alter those genres and discourses to make them more open to feminist models of scientific practice. The goal is not to *feminize* scientific discourse or genres, necessarily, but to work toward changes in scientific rhetoric that would encourage *feminist* values of reflexivity or situated knowledge.

The experiences of women scientists during World War II pinpoint several rhetorical factors that may contribute to women scientists' continued subordination in scientific disciplines and their difficulties trying to enact more socially responsible forms of knowledge: discourses of expertise; gender neutrality; objectivity; and technical rationality, opportunity, and genre. Through further research in these areas, scholars in rhetoric may help to suggest strategies to reform scientific genres and discourses in accordance with feminist values.

Regendering Restrictive Discourses

Deeply entrenched discourses of gender neutrality, objectivity, technical rationality, and expertise permeated the scientific cultures in which women scientists worked during World War II. While these discourses enabled some women scientists to advance professionally during the war, their effects were ambivalent. Often, those discourses required women scientists to adopt dominant assumptions and habits that worked against the interests of women scientists, as a group, even when they worked in favor of some women scientists, as individuals.

Further, because these discourses are embedded in genres, changes to scientific discourse alone—such as employing gender-neutral language—will not fix the problem. These discourses functioned performatively, not simply in texts, but in female scientists' behaviors and actions. For instance, appropriate gender-neutral behavior meant working especially hard to get ahead and subscribing to a meritocratic view of science that assumed hard work would bring professional recognition (Rossiter, *Women Scientists in America* I 267). Most importantly, women scientists were generally not permitted to argue that gender was in any way a factor in one's own career. As Evelyn Fox Keller has written, "Throughout this century, the principal strategy employed by women seeking entrance to the world of science has been premised on the repudiation of gender as a significant variable for scientific productivity. The reasons for this strategy are clear enough: experience has demonstrated all too fully that any acknowledgement of gender-based difference was almost invariably employed as a justification for exclusion" ("The Gender/Science System" 236). Indeed, the trope of gender neutrality continues to pervade scientific institutions, operating as a powerful argument against institutional change (Etzkowitz et al. 54). Similarly, discourses of objectivity, technical rationality, and expertise helped women scientists to maintain or gain what professional status they could, but often at the expense of productive conflict, ethical concerns, or transformations in scientific culture itself.

Ultimately, part of a program for rhetorical studies of science must be to continue to critique and reshape scientific genres and the limiting discourses embedded within them. Some inroads have been made in this respect recently; for example, grant proposals for the National Science Foundation ask scientists to identify conflicts of interest and to show that research involving human subjects conforms to federal policy guidelines (National Science Foundation). Although enacting changes in these discourses and genres may be difficult, scholars in the humanities can at least help our students to analyze, appropriate, and reenvision these genres and discourses. In our courses on (feminist) science studies, the history of (women in) science, the rhetoric of science, and scientific and technical communication, we should encourage attention to scientific language and genres and their gendered implications.

To address the myth of gender neutrality, we might share with our students not only examples of gendered metaphors in scientific language, but also a fuller discussion of how scientific genres have been distributed unequally across genders. We might also ask students to consider whether some genres (Grant proposals? Editorials in scientific magazines? Speeches?) offer more possibilities for raising issues of gender than do others. Finally, we might investigate whether feminist sciences have enacted shifts in the genres of scientific research in the years following World War II.

The discourse of objectivity has already been central to feminist science studies and the history of women in science. But as teachers, we might do more to help our students envision and enact alternatives to the discourse of objectivity. In her article on gender in technical communication, Lee Brasseur has recommended that ethnographies, in particular, offer a way of "granting peripheral voices or subordinate cultures a space within organizational cultures" (482). Following Brassuer's suggestions for ethnographic studies in technical communication, we might ask students to investigate how men and women interact in science classrooms, laboratories, or organizations on campus, whether issues of gender arise in research or classroom settings, and/or how those issues might be silenced in written genres.

As Kirsch has argued, we can also call students' attention to strategies women have developed for confronting issues of authority in academic writing; our histories of women in science can provide us with ample sources for such an endeavor (133). Yet, we must also be careful to complicate the discourse of expertise and its potential effects. As feminist teachers and scholars, we should consider whether our own research enacts the kinds of hierarchies we often wish to contest and critique. We might not only consider how to make our research more accessible to a broad audience outside of our own disciplines and communities, but also consider how to involve others as active participants in our research rather than subjects of our research. In our classrooms, we might examine how scientific genres people encounter in their everyday lives (from advertisements for prescription drugs to nutrition guidelines) encourage a discourse of expertise and downplay individuals' own knowledge, experience, and concerns. I would emphasize here that these suggestions benefit all students, not just female students. However, an understanding of issues of authority, gender, and scientific communication might serve women especially well given the still tenuous place many of them hold in scientific disciplines.

Regendering Genres

The assumption that genres are neutral is misleading. Often the rhetoric of science has worked in a kind of explanatory (even celebratory) manner without critiquing these genres and how they contribute to a culture of science that silences or oppresses women and other marginalized groups. Scholars interested in science studies should consider not only how scientific genres shape scientific knowledge, but who is writing which genres, how genres position individuals differently within scientific networks, and how genres constrain consideration of ethical issues.

One angle might be to focus more attention on the rhetoric of scientific memory—that is, the ways in which accounts of scientific discoveries and events are constructed, often in ways that aggrandize a few key figures, often male, and exclude others who contributed. As Nathan Stormer has argued, memory constrains invention, thus, "to study articulations of rhetoric one must study the normative regimes by which different orders are established, maintained, subverted" (275). Studies of the rhetoric of scientific memory can help us to understand how past articulations of rhetoric limit the invention of new forms of scientific rhetoric and practice.

To take up one example, the genres used on the Manhattan Project precluded recognition of the contributions of individual scientists at the lower ranks, like Marshall and Way. Because the documents produced under the Project were classified, much of the research conducted under its auspices could not be published. Further, the rhetoric used to memorialize the atomic bomb was characterized as explicitly masculine, depicting the Manhattan Project as an "obsessive pursuit of history-making power over nature, reinforced by an aggressive attitude towards [*sic*] nature viewed as female, fuelled by an ever-increasing urgency of its practitioners to be the first to achieve scientific paternity, a pursuit characterized by the total or near total exclusion of affective ties that distract from the task at hand, compounded by insufficient regard or concern for the consequences of the ensuing achievements" (Easlea 58). Given the tendency to make heroes out of a few prominent scientists in the discipline as a whole, it may not be surprising that a few men of scientific genius have been credited with this pursuit—i.e., Robert J. Oppenheimer, Enrico Fermi, Leo Szilard. But this practice occludes the ways in which scientific endeavors, especially since World War II, often depend upon the contributions of tens and even hundreds of individuals working together to solve a problem. Women are especially likely to cluster in the lower ranks of scientific enterprises, performing routine or low status work, and are therefore especially *unlikely* to be recognized for their contributions.

More research is needed on the ways in which genres shape not just status and prestige, but epistemology and identity. The genres Thomas favored on the JERS study prevented anthropologists from writing reflectively about their personal connections to their research. By effacing such individual perspectives, Thomas was ultimately able to produce authoritative, "objective" accounts of the Japanese internment camps, but she did so at the expense of critical reflection on the ethics of internment. The structure of the JERS study ensured that the research conducted by individual informants, like Hankey and Tsuchiyama, became the property of the study itself. While Thomas was able to use that information to her own benefit in the two books she published, Hankey and Tsuchiyama were not. Both young anthropologists had gathered significant data that could have been published as original research, yet neither woman was able to do so after the war. As a result, in-depth, detailed studies of the internment camps were lost in favor of the broader, sweeping studies Thomas published.

Scholars might focus more attention on the gendered nature of teamwork in scientific disciplines, how genres support certain forms of collaboration, and how cooperative efforts in scientific fields may disadvantage women and other minority groups. Lydia J. Roberts employed collaborative strategies to develop the RDAs, and these strategies ultimately helped her to ensure their approval. Many other women contributed to the RDAs by providing scientific data and professional opinions.

Yet, because they were developed by committee, through the National Research Council, they were not credited to the individuals (male or female) who developed them. Over time, the RDAs came to be associated with the NRC and nutrition groups, such as the ADA, not with Roberts or anyone else.

Further, all of the women scientists featured in this book benefited, to some extent, from cooperation with others. Sometimes, this collaboration was productive, in the form of help with drafting, research, writing, editing, and so on. Tsuchiyama received help from Nishimoto, who served as her primary informant. Roberts's work on the RDAs relied on an extensive network of collaborators, which extended beyond the Committee on Dietary Allowances to include a number of researchers in the field. In other cases, this collaboration was supportive, in the form of help with child rearing, domestic help, and other kinds of invisible work.[2] For instance, Roberts never married, but lived with her sister, who did much of the cooking and cleaning so that Roberts could pursue a scientific career. When Leona Marshall gave birth to her first son, her mother provided child care so that Marshall could return to work on the Manhattan Project. What has been the case with collaborative work in general—the lack of recognition granted to both productive and supportive efforts—applies especially to women scientists.

Researchers interested in genre might also continue to research the connections between time and genre, following the example of Yates and Orlikowski, Catherine Schryer, Elinor Ochs and Sally Jacoby, Patricia Dunmire and Hunter Stephenson. As Schryer has suggested in her analysis of chronotopes in the scientific article, when we interrogate a genre, we need to consider its relationship to time, not just in terms of past/present/future, but in terms of "a genre's attempt to control time and space" (82). Following Jan Nespor's study of time and space in undergraduate physics and management education, we might examine, specifically, how writing practices and genres produce and organize time and space in different ways, and how those practices position students of different genders, races, ethnic backgrounds, and so on. For example, one might ask how writing practices differ within those disciplines that currently have a high percentage of women, such as biology, as opposed to those that continue to feature few women, such as physics.

Scholars have noted that the culture of science also depends on an intensive time schedule, on arrangements of *chronos* (chronological time) that build a sense of urgency and excitement (*kairos*) (Ochs and Jacoby; Yates and Orlikowski, "Genre Systems: Chronos"). The cycle of credit also constitutes a temporal structure—an arrangement of time that shapes and is shaped by scientists' activities and actions. Wanda J. Orlikowski and JoAnne Yates explain that temporal structures, which include meeting schedules, project deadlines, academic calendars, tenure clocks, and the like, provide a rhythm and organization to everyday activities. For this reason, they become "templates for the timing and rhythm of members' social action within the community" (Orlikowski and Yates 685). Further, genres play an important role in coordinating and reinforcing arrangements of time (Yates and Orlikowski "Genre Systems: Structuring" 14). In scientific disciplines, researchers typically organize their activities around cyclical timelines, in which certain genres (such as conference proposals, papers, and such) are positioned at certain intervals (Ochs and Jacoby). Genres therefore lend a tempo and rhythm to scientific work. However, less research has focused on how temporal structures of science are organized around gender, as well as genres, in ways that might privilege some scientific workers over others.

In scientific fields, women tended to be more highly concentrated within certain temporal stages than others. Their work was often situated either "downstream" or "upstream" from the most appreciated stages of scientific research and publishing. For instance, women have been more likely to work in lower status positions, as technicians or assistants, who generate data used by scientists in scientific articles. Or, they might be positioned later in the publication cycle, as editors or popularizers of scientific information (Neely; Rauch).

Scientific genres help to reproduce a common culture that requires scientists to work extremely long hours, often late at night. The assumption that scientists must work so intensively has also tended to place women at a disadvantage, particularly women with families. In a 1951 article called "The Woman Problem," Boring drew attention to the image of the "professional fanatic," or the psychologist "who lives primarily for his job—he who eats, sleeps, and finds recreation only because he wishes to work better" (680). Boring concluded that women, whether by desire or necessity, were less likely to assume such a role. The long hours expected of a research scientist conflicted with the temporal structures of child rearing, domestic work, and their spouses' careers (Schiebinger, *Has Feminism* 93–99). Because domestic spaces are considered separate from scientific spaces, women's unequal commitments to the household are considered an inevitable barrier to their advancement—an opinion echoed by former Harvard president Lawrence Summers nearly 50 years later. Yet even those women who are willing to accept long working hours may not feel comfortable doing so. For instance, a female biochemist I interviewed told me that she felt uncomfortable going to the laboratory late at night because she was worried about her personal safety (Rodriguez). Others, like Leona Marshall, may be willing to take on long hours, but not without sacrificing other considerations, such as time with family.

Overall, these examples suggest that, in many scientific disciplines, genres are gendered. This is not to say that scientific genres are inherently antithetical to women, but simply that the ways in which women have historically been positioned within scientific disciplines has also affected the nature of genres. As Debra Journet has argued, genres are shaped not only by specific discursive or textual forms, but also by "the common problems or representations of reality; the preferred methods and techniques; and the whole range of theoretical, methodological, and epistemological commitments that constitutes a discipline" (101). In other words, genres reflect disciplinary and institutional values. The range of rhetorical options available to any individual, then, includes not only vocabulary or textual forms, but also this wider range of representations, methods, theories, and commitments. And all of these factors are determined, in part, by the particular position that an individual occupies within an institutional, disciplinary, and rhetorical context. Further, disciplines tend to privilege genres that preserve and reproduce power relationships so that, in many cases, genres are allocated only to certain individuals, who may use them only in certain ways (115). Such has been the case for women scientists.

Of course, changing scientific genres is no easy task. As Devitt argues, genres develop in the first place because "they respond appropriately to situations that writers encounter repeatedly" (576). Once a genre has become commonplace as a response to such situations, once generic conventions are established, they become increasingly difficult to shift or dislodge. However, this is not to say that only one genre (or genre system) is appropriate for a given rhetorical problem. For instance, the genres that evolved under the Manhattan Project certainly helped to bring the project to completion, and quickly. Yet, the model of the Manhattan Project also assumes a hierarchical organization, involving government, science, and industry, and a single-minded intensity that militates against consideration of social and ethical issues. Alternative models do exist. In their article describing the Human Genome Project, Francis S. Collins, Michael Morgan, and Aristedes Patrinos argue that a more horizontal, participatory approach can be equally effective (289). Whether this "bottom-up" approach led to differences in the genres used within the Human Genome Project might be an interesting line of research. Even in the case of the Manhattan Project, where concerns about speed and secrecy led officials to adopt a top-down system, scientists complained about the compartmentalization of tasks and the curtailment of discussion between scientists. Yet the more open, collaborative model preferred by scientists might have been equally efficient, because it would have allowed for the open exchange of information and ideas.

Other disciplines have witnessed greater changes in the time that has passed since World War II. In anthropology, for instance, scholars now recognize that the conventions of "realist ethnography" failed to allow for the ways in which the researcher's knowledge is shaped by interactions with those she observes, and by the researchers' own attitudes and suppositions (Hansen 176). More research on genres in other contexts might help to identify arrangements that are particularly effective for women scientists and others who have traditionally been marginalized within scientific disciplines.

Regendering Opportunity

In 1944, Margaret Barnard Pickel made the following pronouncement in an article published in the *New York Times:* "We should not take a temporary condition as a guarantee for the future, nor should we take an exception and make a rule from it" (Pickel SM19). In the early years of World War II, it had seemed clear that the war presented some kind of opportunity for women in science. Even before the war ended, however, it became apparent that, for the most part, those opportunities emerged in already gendered scientific niches—in feminized fields, like home economics, or in the less prestigious areas of other disciplines, or in applied work that extended women's existing authority over domestic concerns. This was true not only in home economics and nutrition, which capitalized on women's traditional jurisdiction over food preparation, but also in psychology, where women focused on community outreach work. Even in physics, women were largely confined to performing repetitive, deskilled calculations work, similar to the kind of work deemed particularly suited for women working in war factories. Only a relatively few women—those who were already well positioned in their fields, such as Dorothy Thomas, Lydia Roberts, or Florence Goodenough—were able to direct major wartime projects. Because of their status and power, they were able to "seize the moment" presented by the war. This was especially true in nutrition and home economics, which had already been gendered feminine. For the most part, however, women were systematically channeled into lower status positions.

The nature of women's wartime opportunity did not extend to explicit feminist action. In other words, it did not seem appropriate for women to address the inequalities that obtained in their disciplines. Prominent figures such as Goodenough determined (in part due to the influence of their peers) that the time was not right to lobby for more rigorous changes in their disciplines. Instead, it seemed appropriate to continue to advocate a gender-neutral, meritocratic approach. What is more, once the exigence of war was removed, so too was the sense of urgency that had justified training more women in scientific and technical fields. In fact, a number of postwar factors, such as the G.I. Bill, actually increased opportunities in scientific fields for men.[3] The influx of men into the academy after the war left less space for women. During the war and afterward, opportunities did not open equally to men and women.

However, within American culture in general, and scientific culture in particular, opportunity is theorized without particular attention to gender. Indeed, a central tenet of American culture is that everyone has the opportunity to be successful, so long as they work hard. Yet, women have historically been excluded from socially recognized occasions, forums, and roles necessary to respond to moments of opportunity, especially in science. From the start of the "Scientific Revolution," the laboratory was identified as a masculine space, and scientific rhetoric was similarly defined as a masculine practice (Shapin and Schaffer 65–69; Schiebinger *The Mind Has No Sex?* 158–59). It was more acceptable for women to act as popularizers of science than as practitioners. Women were often excluded from both universities and scientific societies; accordingly, opportunities for women to create new knowledge were limited. Instead, women sought out opportunities to collect information to be used by others, or to popularize knowledge created by others.

Even when the formal barriers to women's entry into the laboratory were removed, a culture of gender neutrality militated against women's ability to foreground gender discrimination. Women's wartime contributions were often limited to the local community—child-care centers, schools, libraries, and so on—rather than the more prestigious government or military spheres. Configurations of both physical and rhetorical space, then, condition the ways in which female rhetors are able to respond to moments of opportunity.

Future research may help to address how this sense of opportunity figured into the rhetorical strategies women scientists used during the conservative environment of the 1950s and 1960s and, later, given the new insights that arose during the women's movement in the 1960s and 1970s. What happened in the 1950s, the period that seemed to mark a low point for women scientists? How did the women's movement change the ways in which women scientists were able to employ scientific rhetoric?

Further, more research is needed on the rhetoric of feminist science that has emerged in the last twenty years. Historians and philosophers of science, such as Nancy Tuana and Londa Schiebinger, have done much to address how feminism has changed the philosophy and history of science, respectively. Comparatively little work has emerged that focuses attention specifically on the rhetorical shifts that may be required to enact a feminist scientific agenda.

More broadly, how have women's material and social positions in different times and places contributed to their ability to respond to new opportunities? More importantly, what kinds of conditions seem to provide the most opportunities for women to participate in rhetorical acts, and how might those conditions be encouraged? A feminist theory of opportunity might focus not simply on identifying and seizing moments, but on strategies that might help to open up opportunities for others.

Too often, women scientists have found that the rhetoric of science has limited their potential to transform their disciplines, even when it ensured their success. As was the case with Marie Curie, for the women featured in this book, their "wounds came from the same source as [their] power." Too often, success meant that their critical insights have remained unsaid, their contributions forgotten. As scholars continue to address the "woman problem in science," I hope that this book will serve as a reminder that change in scientific cultures will rely fundamentally on changes in scientific rhetoric. Creating a more "sustainable science," one that encourages reflexivity and ethical engagement, will require a new vocabulary, a new rhetoric, that will allow scientists to speak and write in a new voice.

NOTES

Introduction

1. For definitions of discourse, see Fairclough (3). Like Fairclough, I take "discourse" to refer to specific examples of interactions between writers and readers, but also to the ways in which symbols (language, images, and such) structure knowledge and practices (3). For Fairclough, "Any discourse is seen as being simultaneously a piece of text, an instance of discursive practice, and an instance of social practice" (4).

Chapter 1. Women Psychologists Forecast Opportunity

Epigraph. From Schwesinger 298.

1. The committee members were drawn from the leadership of the APA and the AAAP, and few women received such appointments, even if they were well qualified. Women represented a disproportionately small number of officials within both of these organizations. In 1943, for instance, women constituted only 10 percent of the official appointees to APA committees, even though they constituted 29 percent of the total number of APA members and associates. Female representation was somewhat better in the AAAP, but still not proportional to the total number of women members (Bryan and Boring, "Women in American Psychology: Prolegomenon" 452). Dr. Alice I. Bryan was later added to the ECP committee in 1943 as the representative of the NCWP (Dallenbach 504).

2. In fact, from the beginning, male psychologists were ambivalent about the role women would play within the discipline, because it was assumed that women's biological roles usurped their professional ones. For instance, G. Stanley Hall, the founder of the American Psychology Association (APA), held that "the woman who eschewed marriage in favor of job concentration selfishly violated her biological ethic—to the detriment of her mammary function, among other evils" (quoted in Russo, "Psychology's Foremothers" 9).

3. In total, the survey identified 3798 psychologists, 2669 male and 1129 female.

4. From their survey, Bryan and Boring found that only 13 of the 247 male respondents were married, a sample they considered too small to be statistically significant, so statistics for unmarried men are not included in the article ("Women in American Psychology: Factors" 4).

5. Terman was perhaps best known for developing the Stanford-Binet Intelligence Test in 1916 and for a series of long-term studies of gifted children, which he conducted in the 1920s.

6. Perhaps the higher employment rates during the war years seemed to indicate success after over a decade of the Depression. However, similar examples of women's scientific organizations disbanding after some initial progress occurred during the Depression. For example, the Association to Aid Scientific Research by Women, founded in 1897, voluntarily disbanded in 1935 because the members determined that women had already achieved equality with their male counterparts (Crawford 493).

Chapter 2. Women Anthropologists Study Japanese Internment

1. Haraway defines "situated knowledges" as a new doctrine of objectivity, one that "privileges contestation, deconstruction, passionate construction, webbed connections, and hope for transformation of systems of knowledge and ways of seeing" (*Simians* 191–92).

2. These documents have been preserved at the Bancroft Library at the University of California, Berkeley.

3. For more on the first (Issei) and second (Nisei) generations of Japanese-Americans, see Wendy Ng in *Japanese American Internment During World War II* (4–6). The Kibei were Japanese-Americans who had been born in the United States, but who returned to Japan for some of their education. The Jun Nisei were Japanese-Americans who had never been to Japan prior to the war.

4. Because the names of evacuees mentioned in JERS archives are protected for reasons of privacy, I will refer to all evacuees by their initials only.

5. Thomas and Hankey also seem to have shared additional disparaging remarks about Tsuchiyama; in one letter, Hankey writes to Thomas: "Was very glad Tamie wasn't in Phoenix. I was in no state to placate 'er" (Hankey to Thomas, 10 February 1944).

6. Thomas did inform Hankey that her materials would be used in *The Spoilage*. In a letter to Hankey, she writes: "We are using your field notes extensively, but we are also using reports and field notes prepared by various other people on the Study to such an extent that it is necessary to put their names on the title page also. Since the only person who has participated with me in the conceptualization and the actual writing of the book is Dick [Nishimoto], he appears as co-author [. . .] In a section called 'Acknowledgments,' reference will, of course, be made to the extensiveness of your field contribution" (Thomas to Hankey, 28 February 1946).

Chapter 3. Women Physicists on the Manhattan Project

Epigraph. Quoted in Sanger 162–63.

1. My definition is a modification of Frank Fischer's description of "technical rationality" in *Technocracy and the Politics of Expertise* (359).

2. In addition to the work undertaken at Los Alamos, Hanford, Oak Ridge, and the University of Chicago, a number of organizations participated, including Westinghouse, Standard Oil, M. W. Kellogg Company, Princeton University, Yale University, Brown University, University of Virginia, Iowa State College, the University of Wisconsin, University of Minnesota, and Columbia University (Stoff, Fanton, and Williams 24–25).

3. Howes and Herzenberg write that women were generally excluded from high-level positions during World War II, with a few exceptions: Mina Rees was head of the mathematics board of the Office of Naval Research, and Gladys Amelia Anslow was chief of Communications of the Information Section of the Office of Field Service of the Office of Scientific Research and Development (OSRD) (18).

4. Secrecy shifted the genre system of scientific research, which normally required that scientists openly share their research through publication. In 1939, British and American physicists voluntarily agreed not to publish on nuclear fission. Later, the military decided that research conducted under the auspices of the Manhattan Project would be heavily classified. At Los Alamos, for instance, documents considered "top-secret" were kept in a locked file, and provisions were in place to encourage scientists and technicians not to leave sensitive documents lying around. (A woman, Charlotte Serber, was appointed scientific librarian in charge of organizing these documents.) Because of the security restrictions, most of the research Marshall and Way performed while working on the Project was never published, but is available now in the form of declassified letters and reports.

5. In a neutron chain reaction, a nucleus fissions, or splits, into two parts. Each part emits, on average, two to three neutrons, each of which, in turn, may cause another nucleus to split, and so on until a continuous reaction occurs (Hawkins 10). This reaction produces large amounts of energy, which scientists hoped to harness to create the bomb.

6. Although Roosevelt authorized research programs on a nuclear weapon in 1940, he did not give priority to the project until 1942, when the Manhattan Engineer District was officially established. In December, 1942, Roosevelt pledged over $500 million to fund the Project (Marceau et al. 1.11).

7. Fermi and Wigner were among a group of over 100 physicists who fled Europe between 1933 and 1941 (Hoddeson et al. 9). In 1935 Wigner was fired from the Institute of Technology in Berlin because his mother was Jewish. He moved to Princeton's Institute for Advanced Study (Sanger 33). Shortly after he won the Nobel Prize in 1938, Fermi escaped from Italy with his wife, Laura, who was Jewish, and his two children.

8. In her study of women who worked as computers in World War II, Jennifer Light writes that "A 'computer' was a human being until approximately 1945. After that date the term referred to a machine, and the former human computers became 'operators'" (469).

9. Marshall writes that "The medical [. . .] were overzealous in watching us for possible radiation effects. They regularly examined our retinas and the bases of our fingernails for signs of tissue damage. At first, they made us have chest X rays every few months, which undoubtedly exposed us to more radiation than we were getting at the pile" (*The Uranium People* 143).

10. Schiebinger provides several similar examples of female scientists who "scheduled" their pregnancies or hid them from their colleagues (*Has Feminism* 95). Schiebinger summarizes that "The goal for these women was to have babies without maternity leave, without a pause in productivity, without appearing to be different from their male colleagues. The result was that they did it at a high cost to themselves and their partners within institutions structured to suppress such things" (95).

11. As Schiebinger writes elsewhere, liberal feminist views assume that "women think and act in ways indistinguishable from men. Only women have babies, but childbirth is supposed to take place exclusively on weekends and holidays, not to disrupt the rhythm of working life. Liberal feminists tend to see sameness and assimilation as the only grounds for equality, and this often requires that women be like men—culturally or even biologically" (*Has Feminism* 3–4).

12. Wende's letters sometimes appear on letterhead from E.I. DuPont de Nemours & Company and his letters are signed "C. J. Wende, Technical Division."

Chapter 4. Women Nutritionists on the NRC

Epigraph. From *Proceedings of the National Nutrition Conference for Defense* 230.

1. For more on the early history and purposes of home economics, see Marjorie M. Brown, *Philosophical Studies of Home Economics in the United States* (East Lansing, MI: Michigan State UP, 1985), and Sarah Stage and Virginia Vincenti, eds., *Rethinking Home Economics* (Ithaca: Cornell University Press, 1997). For more on the rhetoric of pioneers in home economics and nutrition, see Gail Lippincott's "Moving Technical Communication into the Post-Industrial Age: Advice from 1910" (*Technical Communication Quarterly* 12.3 (2003): 321–42) and "Rhetorical Chemistry: Negotiating Gendered Audiences in Nineteenth Century Nutrition Studies" (*Journal of Business and Technical Communication* 14.1 (2003): 10–49).

2. Campbell emphasizes that the "feminine style" is not necessarily one that can be used by women only; men can also employ such a rhetorical style effectively (12). The term *feminine* refers not to any necessary connection to the female sex, but to the fact that female rhetoricians have historically used it because it was conducive to their female experience, their socialized roles (i.e., to expectations about "femininity"), and the rhetorical situations in which they were involved.

3. National Academy of Sciences (NAS)-National Research Council (NRC) Archives: DNRC: Biology and Agriculture: Committee on Food and Nutrition: 1940.

4. National Academy of Sciences (NAS)-National Research Council (NRC) Archives: DNRC: Biology and Agriculture: Committee on Food and Nutrition: Circular Letters: 1941.

5. NAS-NRC Archives: FNB: DNRC: B&A: Committee on Food and Nutrition: Committee on Dietary Allowances: Meetings: Proceedings. Transcript: 2 Jun.

6. In fact, Stiebeling was the first to use the term *dietary allowances* in a 1933 publication on dietary planning for the USDA; she was also first to develop a dietary standard that included vitamins and minerals as well as nutrients such as protein (Harper 3699).

7. As an approximate indication of the probable composition of this audience, in 1938, approximately 3,325 men and 169 women held doctorates in the medical sciences (Rossiter *Women Scientists in America* I 157). Based on this data, women would have represented roughly five percent of medical doctors in 1938. Provided this distribution held true for the Medical Society of the State of New York, women probably constituted only a small portion of the audience for Roberts's "Scientific Basis" speech.

8. As a rough indicator of the composition of this audience, in 1938, women constituted 42.2 percent of scientists in the field of nutrition (Rossiter, *Women Scientists in America* I 136). Yet, for this speech Roberts addressed a group that included not only nutrition experts, but also experts in economics, food policies, agriculture, food production, and international trade. Women were probably less well represented in these latter fields. Presumably, then, in this case Roberts addressed a mixed audience of men and women, with men constituting a majority. Indeed, out of the twenty-three total respondents to Roberts's speech, only one, Reid, was female.

9. In most cases, the RDAs are cited as the product of the NRC's Food and Nutrition Committee, although Fishbein notes that the RDA "was charted by Dr. Lydia Roberts, head of the Department of Home Economics of the University of Chicago" (18).

10. Of the six liaison members from the federal government to the Committee on Food Habits, two were women: Martha M. Eliot, of the Children's Bureau, United States Department of Labor; and Hazel K. Stiebeling, of the Bureau of Home Economics, United States Department of Agriculture (*Proceedings of the National Nutrition Conference for Defense* 254).

11. For more on the domestication of World War II, see Amy Bentley, *Eating for Victory: Food Rationing and the Politics of Domesticity* (Urbana-Champaign: University of Illinois Press, 1998), Nancy A. Walker, *Shaping Our Mother's World: American Women's Magazines.* (Jackson, Miss: University Press of Mississippi, 2000), and Harvey Levenstein, *Paradox of Plenty: A Social History of Eating in Modern America* (Berkeley: University of California Press, 2003).

Conclusion

1. I take the term *regendering* from Cheryl Glenn's *Rhetoric Retold: Regendering the Tradition from Antiquity Through the Renaissance.* Glenn uses "regendering" to characterize her project of identifying how the rhetorical tradition has been shaped by masculine assumptions, recovering contributions of female rhetors, and showing how women's contributions might alter or reshape rhetorical theories and concepts that were defined based on a male standard.

2. For more on productive and supportive models of collaboration, see Lindal Buchanan, "Forging and Firing Thunderbolts: Collaboration and Women's Rhetoric," in *Rhetoric Society Quarterly* 33.4 (2003): 43–63.

3. Passed in 1944, the Servicemen's Readjustment Act (commonly known as the G.I. Bill) included provisions for four years of college education for returning soldiers. Approximately 2.2 million individuals pursued a college or university education under the G.I. Bill, leading to a period of rapid growth of higher education (Loss 889). By 1947–1948, nearly 50 percent of college students in the United States were veterans (889).

BIBLIOGRAPHY

Allison, Samuel. Letter to W. O. Simon. 22 September 1944. Department of Energy Public Reading Room-Hanford Batelle: Richland, Washington.

American Psychological Association. *Publication Manual of the American Psychological Association* 5th ed. Washington, D.C.: American Psychological Association, 2001.

Anthony II, Susan B. *Out of the Kitchen—into the War: Women's Winning Role in the Nation's Drama.* New York: Stephen Daye, Inc., 1943.

Aristotle. *Rhetoric.* Trans. W. Rhys Roberts. New York: The Modern Library, 1954.

Baruch, Dorothy W. "Child Care Centers and the Mental Health of Children in This War." *Journal of Consulting Psychology* 7 (1943): 252–66.

Baym, Nina. *American Women of Letters and the Nineteenth-Century Sciences.* New Brunswick, NJ: Rutgers University Press, 2001.

Bazerman, Charles. "Discursively Structured Activities." *Mind, Culture, and Activity* 4.4 (1997): 296–308.

———. *Shaping Written Knowledge.* Madison: University of Wisconsin Press, 1988.

Bazerman, Charles, Joseph Little, and Teri Chavkin. "The Production of Information for Genred Activity Space." *Written Communication* 20 (2003): 455–77.

Behar, Ruth. "Introduction: Out of Exile." *Women Writing Culture.* Eds. Ruth Behar and Deborah A. Gordon. Berkeley: University of California Press, 1995. 1–29.

———. *The Vulnerable Observer: Anthropology That Breaks Your Heart.* Boston: Beacon, 1996.

Bentley, Amy. *Eating for Victory: Food Rationing and the Politics of Domesticity.* Urbana-Champaign: University of Illinois Press, 1998.

Bergson, Henri. *Time and Free Will: An Essay on the Immediate Data of Consciousness.* Trans. F. L. Pogson. Mineola, NY: Dover, 2001.

Berkenkotter, Carol. "Genre Systems at Work: DSM-IV and Rhetorical Recontextualization in Psychotherapy Paperwork." *Written Communication* 18.3 (2001): 326–49.

Biesecker, Barbara. "Coming to Terms with Recent Attempts to Write Women into the History of Rhetoric." *Philosophy and Rhetoric* 25.2 (1992): 140–61.

Boring, Edwin G. "The Woman Problem." *American Psychologist* 6 (1951): 679–82.

Brasseur, Lee. "Contesting the Objectivist Paradigm: Gender Issues in the Technical and Professional Communication Curriculum." *Central Works in Technical*

Communication. Eds. Johndan Johnson-Eilola and Stuart A. Selber. New York: Oxford University Press, 2004. 475–89.

Brodky, Linda. "Writing Ethnographic Narratives." *Written Communication* 4.1 (1987): 25–48.

Brown, Marjorie M. *Philosophical Studies of Home Economics in the United States: Our Practical-Intellectual Heritage.* Vol.I. East Lansing: College of Human Ecology, Michigan State University, 1985.

Bryan, Alice I. "Educating Civilians for War and Peace through Library Film Forums." *Journal of Consulting Psychology* 7 (1943): 280–88.

Bryan, Alice I., and Edwin G. Boring. "Women in American Psychology: Factors Affecting Their Professional Careers." *American Psychologist* 2.1 (1946): 3–20.

———. "Women in American Psychology: Prolegomenon." *Psychological Bulletin* 41 (1944): 447–54.

———. "Women in American Psychology: Statistics from the OPP Questionnaire." *American Psychologist* 1.3 (1946): 71–79.

Buchanan, Lindal. "Forging and Firing Thunderbolts: Collaboration and Women's Rhetoric." *Rhetoric Society Quarterly* 33 (2003): 43–63.

Bureau of Human Nutrition and Home Economics. *Planning Diets by the New Yardstick of Good Nutrition.* Washington, D.C.: Bureau of Home Economics, U.S. Dept. of Agriculture, 1941.

Burke, Kenneth. *Permanence and Change.* 3d ed. Berkeley: University of California Press, 1984.

Campbell, Karlyn Kohrs. *Man Cannot Speak for Her: A Critical Study of Early Feminist Rhetoric.* Vol.II. New York: Greenwood Press, 1989.

Campbell, Kristen. "The Promise of Feminist Reflexivities: Developing Donna Haraway's Project for Feminist Science Studies." *Hypatia* 19.1 (2004): 161–82.

Capshew, James H., and Alejandra C. Laszlo. "'We Would Not Take No for an Answer': Women Psychologists and Gender Politics During World War II." *Journal of Social Issues* 42.1 (1986): 157–80.

Cautley, Patricia Woodward. "Fifty Years of the International Council of Psychologists." *Psychology in International Perspective: 50 Years of the International Council of Psychologists.* Eds. Uwe P. Gielen, Leonore Loeb Adler, and Norman A. Milgram. Amsterdam: Swets & Zeitlinger, 1992. 3–18.

Ceccarelli, Leah. *Shaping Science with Rhetoric: The Cases of Dobzhansky, Shrödinger, and Wilson.* Chicago: University of Chicago Press, 2001.

Collins, Francis S., Michael Morgan, and Aristedes Patrinos. "The Human Genome Project: Lessons from Large-Scale Biology." *Science* 300 (2003): 286–90.

Committee on Maximizing the Potential of Women in Academic Science and Engineering. *Beyond Bias and Barriers: Fulfilling the Potential of Women in Academic Science and Engineering.* Washington, D.C.: National Academy of Sciences, National Academy of Engineering, and Institute of Medicine, 2006.

Council of Biology Editors. *The CBE Manual for Authors, Editors, and Publishers.* 6th ed. Cambridge: University of Cambridge, 1994.

Crawford, H. Jean. "The Association to Aid Scientific Research by Women." *Science* 76.1978 (1935): 492–93.

Dallenbach, Karl M. "The Emergency Committee in Psychology, National Research Council." *American Journal of Psychology* 59 (1946): 496–582.

de Hoffmann, Frederic. "A Novel Apprenticeship." *All in Our Time: The Reminiscences of Twelve Nuclear Pioneers.* Ed. Jane Wilson. Chicago: The Bulletin of the Atomic Scientists, 1974. 162–73.

Devitt, Amy J. "Generalizing About Genre: New Conceptions of an Old Concept." *College Composition and Communication* 44.4 (1993): 573–86.

Doyle, Margaret D., and Eva D. Wilson. *Lydia Jane Roberts: Nutrition Scientist, Educator, and Humanitarian.* Chicago: The American Dietetic Association, 1989.

"Draft Rejections Surprisingly High." *New York Times* December 1. 1940: 48.

Driskill, Linda. "Understanding the Writing Context in Organizations." *Central Works in Technical Communication.* Eds. Johndan Johnson-Eilola and Stuart A. Selber. New York: Oxford University Press, 2004. 55–69.

"Dr. Stiebeling Gets U.S. Nutrition Post." *New York Times* May 19. 1944: 34.

Dunmire, Patricia L. "Genre as Temporally Situated Social Action: A Study of Temporality and Genre Activity." *Written Communication* 17.1 (2000): 93–138.

Easlea, Brian. *Fathering the Unthinkable: Masculinity, Scientists and the Nuclear Arms Race.* London: Pluto Press, 1983.

"Editorial." *Journal of the American Dietetic Association* 19.2 (1943): 110–11.

"Editorial: The American Dietetic Association and the Defense Program." *Journal of the American Dietetic Association* 17.8 (1941): 790–92.

"Editorial: Preparedness." *Journal of the American Dietetic Association* 16.7 (1940): 683–84.

Eisenhart, Margaret A., and Elizabeth Finkel. *Women's Science: Learning and Succeeding from the Margins.* Chicago: University of Chicago Press, 1998.

Environmental Protection Agency. "Radiation: Risks and Realities". October 8 2004. http://www.epa.gov/rpdweb00/docs/risksandrealities/index.html.

Etzkowitz, Henry, et al. "The Paradox of Critical Mass for Women in Science." *Science* 266.5182 (1994): 51–54.

Fabian, Johannes. *Time and the Work of Anthropology: Critical Essays 1971–1991.* Chur, Switzerland: Harwood, 1991.

Fahnestock, Jeanne. "Accommodating Science: The Rhetorical Life of Scientific Facts." *Written Communication* 15.3 (1998): 330–50.

Fairclough, Norman. *Discourse and Social Change.* Cambridge: Polity Press, 1992.

Feynman, Richard P. "Los Alamos from Below." *Reminiscences of Los Alamos, 1943–1945.* Eds. Lawrence Badash, Joseph O. Hirschfelder, and Herbert P. Broida. Dordrecht: D. Reidel, 1980. 105–32.

Fischer, Frank. *Technocracy and the Politics of Expertise.* Newbury Park: Sage, 1990.

Fishbein, Morris. *The National Nutrition.* Indianapolis: Bobbs-Merrill, 1942.

Fleischman, Suzanne. "Gender, the Personal, and the Voice of Scholarship: A Viewpoint." *Signs* 23.4 (1998): 975–1016.

Gates, Barbara T. *Kindred Nature: Victorian and Edwardian Women Embrace the Living World.* Chicago: University of Chicago Press, 1998.

Gates, Barbara T., and Ann B. Shteir, Eds. *Natural Eloquence: Women Reinscribe Science.* Madison: University of Wisconsin Press, 1997.

Geertz, Clifford. *Works and Lives: The Anthropologist as Author.* Stanford: Stanford University Press, 1988.

Gergen, Kenneth. "Self-Narration in Social Life." *Discourse Theory and Practice: A Reader.* Eds. Margaret Wetherell, Stephanie Taylor, and Simeon J. Yates. London: Sage, 2001. 247–60.

Giangreco, D. M. "Casualty Projections for the U.S. Invasions of Japan, 1945–1946: Planning and Policy Implications." *The Journal of Military History* 61 (1997): 521–82.

Glenn, Cheryl. *Rhetoric Retold: Regendering the Tradition from Antiquity through the Renaissance.* Carbondale: Southern Illinois University Press, 1997.

Gluck, Sherna Berger. *Rosie the Riveter Revisited: Women, the War, and Social Change.* Boston: Twayne Publishers, 1987.

Goodenough, Florence L. "Expanding Opportunities for Women Psychologists in the Post-War Period of Civil and Military Reorganization." *Psychological Bulletin* 41 (1944): 706–12.

———. Letter to Ella Woodward. June 20, 1944. Florence Laura Goodenough Papers, University of Minnesota Archives.

———. Letter to Martha Crumpton Hardy. July 12, 1942. Florence Laura Goodenough Papers, University of Minnesota Archives.

Gross, Alan G. *Starring the Text: The Place of Rhetoric in Science Studies.* Carbondale: Southern Illinois University Press, 2006.

Groves, Leslie R. *Now It Can Be Told: The Story of the Manhattan Project.* New York: Harper and Row, 1962.

Hanfmann, Eugenia. "Eugenia Hanfmann." *Models of Achievement: Reflections of Eminent Women in Psychology.* Eds. Agnes N. O'Connell and Nancy Felipe Russo. Vol.I. New York: Columbia University Press, 1983. 141–52.

Hankey, Rosalie. *Chronological Account of Segregation.* 1943. Japanese American evacuation and resettlement records, BANC MSS 67/14 c, The Bancroft Library, University of California, Berkeley.

———. *Conciliation Begins at Gila.* 1943. Japanese American evacuation and resettlement records, BANC MSS 67/14 c, The Bancroft Library, University of California, Berkeley.

———. *Field Notes.* Murray Wax Papers, Chicago, Newberry Library.

———. Letter to Dorothy Thomas. 12 July 1943. Japanese American evacuation and resettlement records, BANC MSS 67/14c, The Bancroft Library, University of California, Berkeley.

———. Letter to Dorothy Thomas. 17 July 1943. Japanese American evacuation and resettlement records, BANC MSS 67/14c, The Bancroft Library, University of California, Berkeley.

———. Letter to Dorothy Thomas. 25 July 1943. Japanese American evacuation and resettlement records, BANC MSS 67/14c, The Bancroft Library, University of California, Berkeley.

———. Letter to Dorothy Thomas. 19 August 1943. Japanese American evacuation and resettlement records, BANC MSS 67/14c, The Bancroft Library, University of California, Berkeley.

———. Letter to Dorothy Thomas. 19 October 1943. Japanese American evacuation and resettlement records, BANC MSS 67/14c, The Bancroft Library, University of California, Berkeley.

———. Letter to Dorothy Thomas. 20 October 1943. Japanese American evacuation and resettlement records, BANC MSS 67/14c, The Bancroft Library, University of California, Berkeley.

———. Letter to Dorothy Thomas. 1 November 1943. Japanese American evacuation and resettlement records, BANC MSS 67/14c, The Bancroft Library, University of California, Berkeley.

———. Letter to Dorothy Thomas. 15 November 1943. Japanese American evacuation and resettlement records, BANC MSS 67/14c, The Bancroft Library, University of California, Berkeley.

———. Letter to Dorothy Thomas. 17 December 1943. Japanese American evacuation and resettlement records, BANC MSS 67/14c, The Bancroft Library, University of California, Berkeley.

———. Letter to Dorothy Thomas. 1 January 1944. Japanese American evacuation and resettlement records, BANC MSS 67/14c, The Bancroft Library, University of California, Berkeley.

———. Letter to Dorothy Thomas. 10 February 1944. Japanese American evacuation and resettlement records, BANC MSS 67/14c, The Bancroft Library, University of California, Berkeley.

———. Letter to Dorothy Thomas. 26 February 1944. Japanese American evacuation and resettlement records, BANC MSS 67/14c, The Bancroft Library, University of California, Berkeley.

———. Letter to Dorothy Thomas. 5 March 1944. Japanese American evacuation and resettlement records, BANC MSS 67/14c, The Bancroft Library, University of California, Berkeley.

———. *Segregation at Gila.* Japanese American evacuation and resettlement records, BANC MSS 67/14c, The Bancroft Library, University of California, Berkeley.

———. *Synthesis and Additions to Hikada's Report.* Japanese American evacuation and resettlement records, BANC MSS 67/14c, The Bancroft Library, University of California, Berkeley.

———. *Threatened Strike over Reduction in Mess Staff.* Japanese American evacuation and resettlement records, BANC MSS 67/14c, The Bancroft Library, University of California, Berkeley.

Hansen, Kristine. "Rhetoric and Epistemology in the Social Sciences: A Contrast of Two Representative Texts." *Advances in Writing Research, Volume Two: Writing in Academic Disciplines.* Ed. David A. Joliffe. Norwood, NJ: Ablex Publishing Corporation, 1988. 167–210.

Haraway, Donna. *Modest_Witness@Second_Millenium. Femaleman© Meets Oncomouse™.* New York: Routledge, 2007.

———. *Simians, Cyborgs, and Women: The Reinvention of Nature.* New York: Routledge, 1991.

Harper, Alfred E. "Contributions of Women Scientists in the U.S. to the Development of Recommended Dietary Allowances." *The Journal of Nutrition* 133 (2003): 3698–702.

Hawkins, David. *Project Y: The Los Alamos Story.* Los Angeles: Tomash Publishers, 1983.

Hayashi, Brian Masaru. *Democratizing the Enemy: The Japanese American Internment.* Princeton: Princeton University Press, 2004.

Hilgartner, Stephen. *Science on Stage: Expert Advice as Public Drama.* Stanford: Stanford University Press, 2000.

Hirabayashi, Lane Ryo. *The Politics of Fieldwork: Research in an American Concentration Camp.* Tucson: The University of Arizona Press, 1999.

Hoddeson, Lillian, et al. *Critical Assembly: A Technical History of Los Alamos During the Oppenheimer Years, 1943–1945.* Cambridge: Cambridge University Press, 1993.

Hogan, John D., and Virginia Staudt Sexton. "Women and the American Psychological Association." *Psychology of Women Quarterly* 15 (1991): 623–34.

Holden, Constance. "The Mind of the Terrorist." *Science* 307 (2005): 511.

Honey, Maureen. *Creating Rosie the Riveter: Class, Gender, and Propaganda During World War II.* Amherst: University of Massachusetts Press, 1984.

Howes, Ruth H., and Caroline L. Herzenberg. "Chien-Shiung Wu." *Women in Chemistry and Physics: A Bibliographic Sourcebook.* Eds. Louise S. Grinstein, Rose K. Rose, and Miriam H. Rafailovich. Westport, Conn.: Greenwood Press, 1993. 613–19.

———. *Their Day in the Sun: Women of the Manhattan Project.* Philadelphia: Temple University Press, 1999.

Hughes, Jeff. *The Manhattan Project: Big Science and the Atom Bomb.* Revolutions in Science. Ed. Jon Turney. New York: Columbia University Press, 2002.

Hyman, Herbert H. *Taking Society's Measure: A Personal History of Survey Research.* New York: Russell Sage Foundation, 1991.

Institute of Electrical and Electronics Engineers. *IEEE Standards Style Manual.* Institute of Electrical and Electronics Engineers, 2007. Accessed July 18, 2008. http://standards.ieee.org/guides/style/2007_Style_Manual.pdf

Joffe, Natalie F., and Tomannie Thompson Walker. *Some Food Patterns of Negroes in the United States of America and Their Relationship to Wartime Problems of Food and Nutrition.* Washington, D.C.: Committee on Food Habits, National Research Council, 194?.

Journet, Debra. "Writing within (and between) Disciplinary Genres: The 'Adaptive Landscape' as a Case Study in Interdisciplinary Rhetoric." *Post-Process Theory: Beyond the Process Paradigm.* Ed. Thomas Kent. Carbondale: Southern Illinois University Press, 1999. 96–115.

Keller, Evelyn Fox. "The Gender/Science System, or, Is Sex to Gender as Nature Is to Science?" *The Science Studies Reader.* Ed. Mario Biagioli. New York: Routledge, 1999. 234–42.

———. *Reflections on Gender and Science.* New Haven: Yale University Press, 1985.

Kirsch, Gesa. *Women Writing the Academy: Audience, Authority, and Transformation.* Carbondale: Southern Illinois University Press, 1993.

Kohlstedt, Sally Gregory. "Sustaining Gains: Reflections on Women in Science and Technology in 20th-Century United States." *NWSA Journal* 16.1 (2004): 1–26.

Kutz, Eleanor. "Authority and Voice in Student Ethnographic Writing." *Anthropology and Education Quarterly* 21.4 (1990): 340–57.

Latour, Bruno, and Steve Woolgar. *Laboratory Life: The Construction of Scientific Facts.* Princeton, NJ: Princeton University Press, 1986.

Laurence, William L. "Scientists Chart a Diet 'Yardstick' to Give Us Health." *New York Times* May 26. 1941: 1,14.

Lay, Mary M. *The Rhetoric of Midwifery: Gender, Knowledge, and Power.* New Brunswick, NJ: Rutgers University Press, 2000.

Leone, G. E. "Causes for Rejection for Entrance into the Regular Army Due to Physical Defects." *Journal of the American Medical Association* 115 (1940): 1283.

Levenstein, Harvey. *Paradox of Plenty: A Social History of Eating in Modern America* Berkeley: University of California Press, 2003.

Libby, Leona Marshall. *The Uranium People.* New York: Crane, Russak, & Company, Inc., 1979.

Light, Jennifer. "When Computers Were Women." *Technology and Culture* 40.3 (1999): 455–83.

Lippincott, Gail. "Moving Technical Communication into the Post-Industrial Age: Advice from 1910." *Technical Communication Quarterly* 12.3 (2003): 321–42.

———. "Rhetorical Chemistry: Negotiating Gendered Audiences in Nineteenth Century Nutrition Studies." *Journal of Business and Technical Communication* 14.1 (2003): 10–49.

Lomon Koos, Earl. "A Study of the Use of the Friendship Pattern in Nutrition Education." *The Problem of Changing Food Habits: Report of the Committee on Food Habits, 1941–1943.* Eds. Carl Guthe and Margaret Mead. Washington, D.C.: National Research Council, National Academy of Sciences, 1943. 74–81.

Loss, Christopher P. "The Most Wonderful Thing Has Happened to Me in the Army: Psychology, Citizenship, and American Higher Education in World War II." *Journal of American History* 92 (2005): 864–91.

Macy, Icie G., and Harold H. Williams. *Hidden Hunger.* Lancaster, Penn.: The Jaques Cattell Press, 1945.

Manley, J. H. "Organizing a Wartime Laboratory." *All in Our Time: The Reminiscences of Twelve Nuclear Pioneers.* Ed. Jane Wilson. Chicago: The Bulletin of the Atomic Scientists, 1974. 126–41.

Marceau, Thomas E., et al. "History of the Plutonium Production Facilities at the Hanford Site District, 1943–1990." U.S. Department of Energy: Hanford Cultural and Historic Resources Program, 2002.

Marcus, George E. "Rhetoric and the Ethnographic Genre in Anthropological Research." *A Crack in the Mirror: Reflexive Perspectives in Anthropology.* Ed. Jay Ruby. Philadelphia: University of Pennsylvania Press, 1982. 163–71.

Marquis, Donald G. "The Mobilization of Psychologists for War Service." *Psychological Bulletin* 41 (1944): 469–73.

Martin, A. Leila. "Cooperation of the Child Study Department of Rochester Board of Education with Selective Service." *Journal of Consulting Psychology* 7 (1943): 267–79.

Martin, Emily. "The Egg and the Sperm: How Science Has Constructed a Romance Based on Stereotypical Male-Female Roles." *Signs* 16.3 (1991): 485–501.

Martin, Murray J., et al. "Katharine Way." *Women in Chemistry and Physics:*

A Biobibliographic Sourcebook. Eds. Louise S. Grinstein, Rose K. Rose, and Miriam H. Rafailovich. Westport, Conn.: Greenwood Press, 1993. 572–80.

Mead, Margaret. "Dietary Patterns and Food Habits." *Journal of the American Dietetic Association* 19.1 (1943): 1–5.

———. "The Problem of Changing Food Habits." *The Problem of Changing Food Habits: Report of the Committee on Food Habits, 1941–1943.* Eds. Carl E. Guthe and Margaret Mead. Washington, D.C.: National Research Council, National Academy of Sciences, 1943. 20–31.

Milkman, Ruth. *Gender at Work: The Dynamics of Job Segregation by Sex During World War II.* Urbana: University of Illinois Press, 1987.

Miller, Carolyn R. "Genre as Social Action." *Quarterly Journal of Speech* 70 (1984): 151–67.

———. "Kairos in the Rhetoric of Science." *A Rhetoric of Doing: Essays on Written Discourse in Honor of James L. Kinneavy.* Eds. Stephen P. White, Neil Nakadate, and Roger D. Cherry. Carbondale: Southern Illinois University Press, 1992. 310–27.

Mitchell, Helen S. "The National Nutrition Outlook." *Journal of Home Economics* 33 (1941): 537–40.

Myers, Greg. *Writing Biology.* Madison: University of Wisconsin Press, 1990.

National Research Council. *Minutes: Meeting of the Committee on Food and Nutrition.* National Academy of Sciences (NAS)-National Research Council (NRC) Archives: DNRC: Biology and Agriculture: Committee on Food and Nutrition, Washington, D.C.

———. *Proceedings: Committee on Dietary Allowances.* The Master Reporting Company, Inc., 1944. National Academy of Sciences (NAS)-National Research Council (NRC) Archives: DNRC: Biology and Agriculture: Committee on Food and Nutrition, Washington, D.C.

National Research Council, Committee on Food and Nutrition. *Recommended Dietary Allowances.* Washington, D.C.: National Research Council 1941.

National Research Council, Committee on Food Habits. *Manual for the Study of Food Habits.* Bulletin of the National Research Council, Number 111. Washington, D.C.: National Research Council, Academy of Sciences, 1945.

———. *The Problem of Changing Food Habits.* Washington, D.C.: National Research Council, National Academy of Sciences, 1943.

National Science Foundation. *Grant Proposal Guide.* Washington, D.C.: National Science Foundation, 2004.

Neely, Kathryn A. "Woman as Mediatrix: Women as Writers on Science and Technology in the Eighteenth and Nineteenth Centuries." *IEEE Transactions on Professional Communication* 35 (1992): 208–16.

Nerad, Maresi. *The Academic Kitchen: A Social History of Gender Stratification at the University of California, Berkeley.* Albany: State University of New York Press, 1999.

Nespor, Jan. *Knowledge in Motion: Space, Time and Curriculum in Undergraduate Physics and Management.* London: The Farmer Press, 1994.

Newman, Louise M. "Coming of Age, but Not in Samoa: Reflections on Margaret Mead's Legacy for Western Liberal Feminism." *American Quarterly* 48.2 (1996): 233–72.

Ng, Wendy. *Japanese American Internment During World War II: A History and Reference Guide.* Westport, Conn.: Greenwood Press, 2002.

Nye, Robert A. "Medicine and Science as Masculine 'Fields of Honor.'" *OSIRIS* 12 (1997): 60–79.

Ochs, Elinor, and Sally Jacoby. "Down to the Wire: The Cultural Clock of Physicists and the Discourse of Consensus." *Language in Society* 26.4 (1997): 479–505.

Oldenziel, Ruth. *Making Technology Masculine: Men, Women and Modern Machines in America, 1870–1945.* Amsterdam: Amsterdam University Press, 1999.

Orlikowski, Wanda, and JoAnne Yates. "It's About Time: Temporal Structuring in Organizations." *Organization Science* 13.6 (2002): 684–700.

Paré, Anthony. "Discourse Regulations and the Production of Knowledge." *Writing in the Workplace: New Research Perspectives.* Ed. Rachel Spilka. Carbondale: Southern Illinois University Press, 1993. 111–23.

———. "Genre and Identity: Individuals, Institutions, and Ideology." *The Rhetoric and Ideology of Genre.* Eds. Richard Coe, Lorelei Lingard, and Tatiana Teslenko. Creskill, NJ: Hampton Press, 2002. 57–71.

Pattee, Alida Frances. *Vitamins and Minerals for Everyone.* New York: G.P. Putnam's Sons, 1942.

Pickel, Margaret Barnard. "A Warning to the Career Woman." *New York Times* 1944: SM19.

Proceedings of the National Nutrition Conference for Defense. National Nutrition Conference for Defense. 1941. United States Government Printing Office.

"Rationing of Meats, Fish, Fats, Oils, and Cheese Begins." *Journal of the American Dietetic Association* 19 (1943): 290.

Rauch, Alan. "Mentoria: Women, Children, and the Structures of Science." *Nineteenth Century Contexts* 27.4 (2005): 335–51.

"A Renewed Urgency on AIDS." *New York Times* December 3. 1993: A32.

Rich, Adrienne. "The Antifeminist Woman." *On Lies, Secrets, and Silence.* New York: W. W. Norton, 1979. 69–84.

———. "Power." *The Fact of a Doorframe: Poems Selected and New 1950–1984.* New York: W. W. Norton, 1994. 225.

Roberts, Lydia J. "Beginnings of the Recommended Dietary Allowances." *Journal of the American Dietetic Association* 34 (1958): 903–8.

———. *Circular Letter No.2.* National Academy of Sciences (NAS)-National Research Council (NRC) Archives: DNRC: Biology and Agriculture: Committee on Food and Nutrition: Circular Letters: 1941.

———. "Improvement of the Nutritional Status of American People." *Journal of Home Economics* 36 (1944): 401–4.

———. *Nutrition Work with Children.* Chicago: University of Chicago Press, 1927.

———. "Scientific Basis for the Recommended Dietary Allowances." *Journal of the American Dietetic Association* (1944): 59–66.

———. "The Usefulness and Validity of the Dietary Allowances." *Food for the World.* Presented at the Norman Watt Harris Memorial Foundation's Twentieth Institute, Chicago, September 4–8, 1944. Ed. Theodore W. Schultz. Chicago: University of Chicago Press, 1945. 105–14.

Rodriguez, Maria. Personal Interview. 2006.

Rogers, Priscilla S., and Song Mei Lee-Wong. "Reconceptualizing Politeness to Accommodate Dynamic Tensions in Subordinate-to-Superior Reporting." *Journal of Business and Technical Communication* 17.4 (2003): 379–412.

Roosevelt, Franklin D. "President's Address Asking More Funds for Defense." *New York Times* July 11. 1940: 10.

Rossiter, Margaret W. *Women Scientists in America: Before Affirmative Action, 1940–1972.* Vol.II. Baltimore: Johns Hopkins University Press, 1995.

———. *Women Scientists in America: Struggles and Strategies to 1940.* Vol.I. Baltimore: Johns Hopkins University Press, 1982.

Rude, Carolyn D. "The Report for Decision Making: Genre and Inquiry." *Central Works in Technical Communication.* Eds. Johndan Johnson-Eilola and Stuart A. Selber. New York: Oxford University Press, 2004. 70–90.

Russo, Nancy Felipe. "Psychology's Foremothers: Their Achievements in Context." *Models of Achievement: Reflections of Eminent Women in Psychology.* Eds. Agnes N. O'Connell and Nancy Felipe Russo. New York: Columbia University Press, 1983. 9–24.

Russo, Nancy Felipe, and Agnes N. O'Connell. "Models from Our Past: Psychology's Foremothers." *Psychology of Women Quarterly* 5 (1980): 11–54.

Sanger, S. L. *Working on the Bomb: An Oral History of WWII Hanford.* Ed. Craig Wollner. Portland: Portland State University Press, 1995.

Schiebinger, Londa. "Creating Sustainable Science." *OSIRIS* 12 (1997): 201–16.

———. *Has Feminism Changed Science?* Cambridge, Mass.: Harvard University Press, 1999.

———. *The Mind Has No Sex? Women in the Origins of Modern Science.* Cambridge, Mass.: Harvard University Press, 1989.

———. *Nature's Body: Gender in the Making of Modern Science.* Boston: Beacon, 1993.

Schryer, Catherine F. "Genre Time/Space: Chronotopic Strategies in the Experimental Article." *JAC* 19 (1999): 81–89.

Schultz, Theodore W. "Introduction." *Food for the World.* Ed. Theodore W. Schultz. Chicago: University of Chicago Press, 1944. v–xi.

Schwesinger, Gladys C. "The National Council of Women Psychologists." *Journal of Consulting Psychology* 7 (1943): 298–301.

Service, Robert F. "DOE Pushes for Solar Power." *Science* 308 (2005): 1391.

Seward, Georgene H. "Sex Roles in Postwar Planning." *The Journal of Social Psychology* 19 (1944): 163–85.

Shapin, Steven, and Simon Schaffer. *Leviathan and the Air-Pump: Hobbes, Boyle and the Experimental Life.* Princeton: Princeton University Press, 1985.

Sheffield, Suzanne Le-May. *Revealing New Worlds: Three Victorian Women Naturalists.* London: Routledge, 2001.

Stage, Sarah, and Virginia Vincenti, Eds. *Rethinking Home Economics.* Ithaca: Cornell University Press, 1997.

Starn, Orin. "Engineering Internment: Anthropologists and the War Relocation Authority." *American Ethnologist* 13.4 (1986): 700–20.

Steele, Evelyn. *Wartime Opportunities for Women.* New York: E.P. Dutton & Co., Inc., 1943.

Steele, Karen Dorn. "Tracking Down Hanford's Victims." *Bulletin of the Atomic Scientists* 46.8 (1990): *n.p.*

Stephenson, Hunter W. *Forecasting Opportunity: Kairos, Production, and Writing.* Lanham: University Press of America, 2005.

Stoff, Michael B., Jonathan F. Fanton, and R. Hal Williams, Eds. *The Manhattan Project: A Documentary Introduction to the Atomic Age.* New York: McGraw-Hill Book Company, Inc., 1991.

Stone, R. S. *Exposure Exceeding Tolerance.* Richland, Wash. E I. Du Pont de Nemours & Company, Inc., 1944.

Stormer, Nathan. "Articulation: A Working Paper on Rhetoric and Taxis." *Quarterly Journal of Speech* 90.3 (2004): 257–84.

Stratton, Dorothy C., and Doris C. Springer. "Problems of Procurement, Training, and Morale among Members of the Women's Reserve of the United States Coast Guard Reserve." *Journal of Consulting Psychology* 7 (1943): 274–79.

Summers, Lawrence. *Remarks at NBER Conference on Diversifying the Science & Engineering Workforce.* Cambridge, Mass.: Harvard University Press, 2005.

Szilard, Leo. "A Petition to the President of the United States." U.S. National Archives, Record Group 77, Records of the Chief of Engineers, Manhattan Engineer District, Harrison-Bundy File, Folder #76.: Washington, D.C.

Terman, Lewis. Letter to Florence Goodenough. June 27, 1942. Florence Laura Goodenough papers, University of Minnesota Archives.

Thomas, Dorothy Swaine. "Experiences in Interdisciplinary Research." *American Sociological Review* 17.6 (1952): 663–69.

———. Letter to Rosalie Hankey. 24 July 1943. Japanese American evacuation and resettlement records, BANC MSS 67/14 c, The Bancroft Library, University of California, Berkeley.

———. Letter to Rosalie Hankey. 1 September 1943. Japanese American evacuation and resettlement records, BANC MSS 67/14 c, The Bancroft Library, University of California, Berkeley.

———. Letter to Rosalie Hankey. 10 November 1943. Japanese American evacuation and resettlement records, BANC MSS 67/14 c, The Bancroft Library, University of California, Berkeley.

———. Letter to Rosalie Hankey. 15 November 1943. Japanese American evacuation and resettlement records, BANC MSS 67/14 c, The Bancroft Library, University of California, Berkeley.

———. Letter to Rosalie Hankey. 30 December 1943. Japanese American evacuation and resettlement records, BANC MSS 67/14 c, The Bancroft Library, University of California, Berkeley.

———. Letter to Rosalie Hankey. Undated. Japanese American evacuation and resettlement records, BANC MSS 67/14 c, The Bancroft Library, University of California, Berkeley.

———. Letter to Tamie Tsuchiyama. 15 July 1944. Japanese American evacuation and resettlement records, BANC MSS 67/14 c, The Bancroft Library, University of California, Berkeley.

———. Letter to Tamie Tsuchiyama. 26 January 1944. Japanese American evacuation and resettlement records, BANC MSS 67/14 c, The Bancroft Library, University of California, Berkeley.

———. Letter to Tamie Tsuchiyama. 28 July 1944. Japanese American evacuation and resettlement records, BANC MSS 67/14 c, The Bancroft Library, University of California, Berkeley.

———. Letter to Tamie Tsuchiyama. 2 August 1944. Japanese American evacuation and resettlement records, BANC MSS 67/14 c, The Bancroft Library, University of California, Berkeley.

Thomas, Dorothy Swaine, Charles Kikuchi, and James Sakoda. *The Salvage.* Berkeley: University of California Press, 1952.

Thomas, Dorothy Swaine, and Richard S. Nishimoto. *The Spoilage.* Berkeley: University of California Press, 1946.

Thompson, Dennis N. "Florence Laura Goodenough." *Women in Psychology: A Bio-Bibliographic Sourcebook.* Eds. Agnes N. O'Connell and Nancy Felipe Russo. Westport, Conn.: Greenwood Press, 1990. 125–35.

Thornton, Robert J. "The Rhetoric of Ethnographic Holism." *Cultural Anthropology* 3.3 (1988): 285–303.

Tolman, Ruth S. "Wartime Organizational Activities of Women Psychologists: I. Subcommittee of the Emergency Committee of the Services of Women Psychologists." *Journal of Consulting Psychology* 7.6 (1943): 296–97.

Tsuchiyama, Tamie. *The Beating of S______ K______.* 1943. Japanese American evacuation and resettlement records, BANC MSS 67/14 c, The Bancroft Library, University of California, Berkeley.

———. *Chronological Account of the Poston Strike.* 1942. Japanese American evacuation and resettlement records, BANC MSS 67/14 c, The Bancroft Library, University of California, Berkeley.

———. *History of the Central Executive Committee.* 1943. Japanese American evacuation and resettlement records, BANC MSS 67/14 c, The Bancroft Library, University of California, Berkeley.

———. Letter to Dorothy Thomas. 24 August 1942. Japanese American evacuation and resettlement records, BANC MSS 67/14 c, The Bancroft Library, University of California, Berkeley.

———. Letter to Dorothy Thomas. 19 November 1943. Japanese American evacuation and resettlement records, BANC MSS 67/14 c, The Bancroft Library, University of California, Berkeley.

———. Letter to Dorothy Thomas. 3 January 1944. Japanese American evacuation and resettlement records, BANC MSS 67/14 c, The Bancroft Library, University of California, Berkeley.

———. Letter to Dorothy Thomas. 24 January 1944. Japanese American evacuation and resettlement records, BANC MSS 67/14 c, The Bancroft Library, University of California, Berkeley.

———. Letter to Dorothy Thomas. 31 January 1944. Japanese American evacuation and resettlement records, BANC MSS 67/14 c, The Bancroft Library, University of California, Berkeley.

———. Letter to Dorothy Thomas. 12 July 1944. Japanese American evacuation and resettlement records, BANC MSS 67/14 c, The Bancroft Library, University of California, Berkeley.

———. Letter to Dorothy Thomas. 17 July 1944. Japanese American evacuation and resettlement records, BANC MSS 67/14 c, The Bancroft Library, University of California, Berkeley.

———. *Notes on Selective Service Registration.* 1943. Japanese American evacuation and resettlement records, BANC MSS 67/14 c, The Bancroft Library, University of California, Berkeley.

Tuana, Nancy. "The Values of Science: Empiricism From a Feminist Perspective." *Synthese* 104.3 (1995): 1–21.

United States Census Bureau. *Statistical Abstract of the United States: 20th Century Statistical Trends.* Washington, D.C., 1999.

United States Civil Service Commission. *Calling Women for Federal War Work.* Washington, D.C.: United States Government Printing Office, 1943.

Walker, Nancy A. *Shaping Our Mother's World: American Women's Magazines.* Jackson, Miss.: University Press of Mississippi, 2000.

Walsh, Mary Roth. "Academic Professional Women Organizing for Change: The Struggle in Psychology." *Journal of Social Issues* 41.4 (1985): 17–28.

"Warns of Shortage of Fats in Our Diet." *New York Times* October 24. 1942: 12.

Watts, Ruth. *Women in Science: A Social and Cultural History.* London: Routledge, 2007.

Wax, Rosalie Hankey. *Doing Fieldwork: Warnings and Advice.* Chicago: University of Chicago Press, 1971.

———. "Twelve Years Later: An Analysis of Field Experience." *The American Journal of Sociology* 63.2 (1957): 133–42.

Way, Katharine. "Letter to Charles Wende." 8 July 1944. Department of Energy Public Reading Room-Hanford Batelle: Richland, Washington.

———. "Memorandum to Samuel Allison." 11 August 1944. Department of Energy Public Reading Room-Hanford Batelle: Richland, Washington.

———. *Time Interval between Discharge of Metal from the Pile and Processing in Canyon.* Hanford: Hanford Engineer Works, 1944.

Wells, Susan. *Out of the Dead House: Nineteenth-Century Women Physicians and the Writing of Medicine.* Madison: University of Wisconsin Press, 2001.

Wende, Charles. Letter to Katharine Way. 1 August 1944. Department of Energy Public Reading Room-Hanford Batelle: Richland, Washington.

———. Letter to Samuel Allison. 1 September 1944. Department of Energy Public Reading Room-Hanford Batelle: Richland, Washington.

Wheeler, John A. *Limiting Mass II.* Hanford: Hanford Engineer Works, 1945.

Wheeler, John A., and L. W. Marshall. *Limiting Mass.* Richland, WA: E I. Du Pont de Nemours & Company, Inc., 1945.

———. *Limiting Mass: An Application and Consideration Presented in a Memorandum of the Same Title, March 29, 1945.* Hanford: Hanford Engineer Works, 1945.

Wheeler, John Archibald, and Kenneth Ford. *Geons, Black Holes, and Quantum Foam: A Life in Physics.* New York: W. W. Norton, 1998.

Wilder, Russell M. "Nutrition and National Defense." *Journal of the American Dietetic Association* 18.1 (1942): 1–8.

Wilmot, Jennie S. "Substitutes, Extenders, Deceivers." *Journal of the American Dietetic Association* 19.6 (1943): 505–6.

Wilson, M. L. "Nutrition and Defense." *Journal of the American Dietetic Association* 17.1 (1941): 12–20.

Winner, Langdon. *The Whale and the Reactor: A Search for Limits in an Age of High Technology.* Chicago: University of Chicago Press, 1986.

Winsor, Dorothy. *Writing Power: Communication in an Engineering Center.* Albany: State University of New York Press, 2002.

Woolf, S. J. "Chief of Staff on the Science Front." *New York Times* 1944: SM16.

Yates, JoAnne, and Wanda Orlikowski. "Genre Systems: Chronos and Kairos in Communicative Interaction." *The Rhetoric and Ideology of Genre.* Eds. Richard Coe, Lorelei Lingard, and Tatiana Teslenko. Creskill, NJ: Hampton Press, 2002. 103–21.

———. "Genre Systems: Structuring Interaction through Communication Norms." *Journal of Business Communication* 39.1 (2002): 13–35.

INDEX

Page references in italics refer to illustrations.

JORDYNN JACK is assistant professor of English at the University of North Carolina at Chapel Hill, where she teaches courses in rhetoric of science, feminist rhetoric, and science writing. She has published articles in such journals as *College English*, *Rhetoric Society Quarterly*, and the *Quarterly Journal of Speech*, and has an essay in the edited collection *Webbing Cyberfeminist Practice: Communities, Pedagogies, and Social Action*.

The University of Illinois Press is a founding member of the Association of American University Presses.

Composed in 9.5/12.5 Trump Mediaeval by Jim Proefrock at the University of Illinois Press
Manufactured by Cushing-Malloy, Inc.

University of Illinois Press
1325 South Oak Street
Champaign, IL 61820-6903
www.press.uillinois.edu